ETHICS AND INTEGRITY IN VISUAL RESEARCH METHODS

ADVANCES IN RESEARCH ETHICS AND INTEGRITY

Series Editor: Dr Ron Iphofen, *FAcSS, Independent Consultant, France*

Recent Volumes

Volume 3: Virtue Ethics in the Conduct and Governance of Social Science Research, *Edited by Nathan Emmerich*
Volume 4: Ethics and Integrity in Health and Life Sciences Research, *Edited by Zvonimir Koporc*

ADVANCES IN RESEARCH ETHICS AND INTEGRITY, VOLUME 5

ETHICS AND INTEGRITY IN VISUAL RESEARCH METHODS

EDITED BY

SAVANNAH DODD
Queen's University Belfast, UK

emerald PUBLISHING

United Kingdom – North America – Japan
India – Malaysia – China

Emerald Publishing Limited
Howard House, Wagon Lane, Bingley BD16 1WA, UK

First edition 2020

Reprints and permissions service
Contact: permissions@emeraldinsight.com

British Library Cataloguing in Publication Data
A catalogue record for this book is available from the British Library

ISBN: 978-1-78769-420-0 (Print)
ISBN: 978-1-78769-419-4 (Online)
ISBN: 978-1-78769-421-7 (Epub)

ISSN: 2398-6018 (Series)

Printed and bound by CPI Group (UK) Ltd, Croydon, CR0 4YY

ISOQAR certified Management System, awarded to Emerald for adherence to Environmental standard ISO 14001:2004.

ISOQAR REGISTERED
Certificate Number 1985
ISO 14001

INVESTOR IN PEOPLE

CONTENTS

LIST OF FIGURES AND TABLES

ABOUT THE SERIES EDITOR

Dr Ron Iphofen, FAcSS, is Executive Editor of the Emerald book series *Advances in Research Ethics and Integrity* and edited volume 1 in the series, *Finding Common Ground: Consensus in Research Ethics Across the Social Sciences* (2017). He is an Independent Research Consultant, a Fellow of the UK Academy of Social Sciences, the Higher Education Academy and the Royal Society of Medicine. Since retiring as Director of Postgraduate Studies in the School of Healthcare Sciences, Bangor University, his major activity has been as an Adviser to the European Commission (EC) and its agencies, the European Research Council and the Research Executive Agency on both the Seventh Framework Programme and Horizon 2020. His consultancy work has covered a range of research agencies (in government and independent) across Europe. He was Vice Chair of the UK Social Research Association (SRA), updated their Ethics Guidelines and now convenes the SRA's Research Ethics Forum. He was Scientific Consultant for the EC RESPECT project – establishing pan-European standards in the social sciences and chaired the Ethics and Societal Impact Advisory Group for another EC-funded European Demonstration Project on mass transit security (SECUR-ED). He has advised the UK Research Integrity Office, the National Disability Authority of the Irish Ministry of Justice, the UK Parliamentary Office of Science and Technology, the Scottish Executive, UK Government Social Research, National Centre for Social Research, the Audit Commission, the Food Standards Agency, the Ministry of Justice, the BIG Lottery, a UK Local Authorities' Consortium, Skills Development Scotland and Agence Nationale de la Recherche (ANR the French Research Funding agency) among many others. Ron was founding Executive Editor of the Emerald gerontology journal *Quality in Ageing and Older Adults*. He published *Ethical Decision Making in Social Research: A Practical Guide* (Palgrave Macmillan, 2009 and 2011) and coedited with Martin Tolich *The SAGE Handbook of Qualitative Research Ethics* (Sage, 2018). He is currently leading a new €2.8M EC-funded project (PRO-RES) that aims at promoting ethics and integrity in all non-medical research (2018–2021).

ABOUT THE CONTRIBUTORS

Susan M. Brigham is Professor in the Faculty of Education at Mount Saint Vincent University, Chair of the Alexa McDonough Institute for Women, Gender and Social Justice, and President of the Canadian Association for the Study of Adult Education. Her research interests include adult education, arts-based research methods, migration, and diversity issues. Her co-edited books include *Adult Education and Learning in Canada* (Thompson, forthcoming 2020); *Building on Critical Traditions: Adult Education and Learning in Canada* (Thompson, 2013), and *Africentricity in Action* (Fernwood, 2012). She is Associate Editor of the *Canadian Journal for the Study of Adult Education* and the *Alberta Journal of Educational Research*. Susan has conducted research and presented her work in Africa, Asia, the Caribbean, Europe, and North America.

Jennie Carlsten has published on Irish cinema, documentary, cinematic emotion, and film and history. Her research explores the role of emotion in film viewing, looking at formal strategies and audience response within the context of emotion theories. She is the co-editor of the volume *Film, History and Memory*. Dr Carlsten is a Lecturer in the School of Communication and Media at Ulster University in Northern Ireland. She also works for Film Hub NI, where she is researching the UK's community cinemas.

Ciara Chambers is Head of the Department of Film and Screen Media, University College Cork. She is the Author of *Ireland in the Newsreels* (Irish Academic Press, 2012), Co-editor of *Researching Newsreels: Local, National and Transnational Case Studies* (Palgrave, 2018), and Screenwriter and Associate Producer of the six-part television series *Éire na Nuachtscannán* (TG4, 2017). She is a member of the Board of Irish Screen Studies, the Council of the International Association of Media and History, and the Northern Ireland Film Heritage and Archive Working Group. She is Associate Editor of *The Historical Journal of Film, Radio and Television* and a member of the editorial teams of *Estudios Irlandeses* and *Alphaville: Journal of Film and Screen Media*. She has worked on a range of archival projects and digitisation initiatives with the Irish Film Archive, Northern Ireland Screen, Belfast Exposed Photography, UTV, BBC, and the British Universities Film and Video Council (now Learning on Screen).

Savannah Dodd is an Anthropologist and a Documentary Photographer based in Belfast. Savannah integrated her knowledge of anthropology with her passion for photography in 2017 when she founded the Photography Ethics Centre with the aim of raising awareness about ethics and increasing ethical literacy across the photography industry. Her work with the Centre involves developing online training courses, designing and facilitating workshops, authoring articles for online and print publications, and lecturing at universities on the topic of photography ethics. She has organised events on photography ethics at institutions like the Royal Photographic Society and PhotoIreland's Museum of Contemporary Photography. Alongside her work with the Centre, she is pursuing her PhD in Anthropology looking at the ethics and politics of archiving photographs in Northern Ireland.

Cigdem Esin is a Senior Lecturer in Psychosocial Studies at the University of East London and a Co-director of the Centre for Narrative Research. Her research focusses on narrative methodologies, multimodal and visual narratives, and research ethics.

Robert Godden is a Human Rights Campaign and Communication Consultant specialising in the use of visual media for positive social change. From 2000 to 2014 he worked in the Asia-Pacific Programme at the International Secretariat of Amnesty International, based in London (2000–2008), Kathmandu (2009–2011), and Hong Kong (2011–2014). Between 2004 and 2014, he was the organisation's Asia-Pacific Campaign Coordinator. In 2014, he co-founded Rights Exposure, a consultancy providing support to non-profits, including research, project management, training, and audio-visual production. In 2018, he collaborated with the Journalism and Media Studies Centre (University of Hong Kong) and the Explore Program (World Press Photo Foundation) to organise the conference, 'The Ethical Image: Challenges in Visualising a Changing World'. The conference covered a wide range of issues, including informed consent and the ethical representation of vulnerable populations. He has a Master's degree in South Asia Area Studies from the School of Oriental and African Studies, University of London.

Crispin Hughes is a Photographer who focusses on documenting social issues in the UK and internationally. He has worked in participatory photography with a wide range of communities, including school children in the UK and communities living with HIV in South Africa.

Mohamed Kharbach is a doctoral candidate at Mount Saint Vincent University. His doctoral research centres on the analysis of the discourse of radicalised groups. His research interests include: critical discourse analysis, critical pedagogy, critical race theory, visual research methodologies, and emerging literacies. He is an Associate Copyeditor in the *Canadian Journal for New Scholars in Education*.

Aura Lounasmaa is a Lecturer in Social Sciences at the University of East London (UEL). She worked as part of the award-winning team from UEL in the Calais Jungle to teach a university course in the camp in 2015–2016. She currently directs the Erasmus+ funded pre-sessional university course Open Learning Initiative for refugees and asylum seekers in the UK.

Angela Stephanie Mazzetti is a Senior Lecturer at Newcastle University Business School. Angela's research focusses on the ways in which stress appraisal and coping behaviours are influenced by cultural norms. Angela has published a number of papers on the ethical and practical challenges of engaging in sensitive qualitative research. Papers related to the use of drawings in sensitive research include Mazzetti and Blenkinsopp (2012) and Mazzetti (2014).

Cahal McLaughlin is Professor of Film Studies at Queen's University Belfast and Director of the Prisons Memory Archive (www.prisonsmemoryarchive.com) and coinvestigator on the AHRC funded project Community Experience of Conflict in Haiti: Assessing the Emotional Legacy of Civilian Deaths as a result of Intense Use of Force by UN Peacekeepers. He has worked on films in South Africa and Ireland, exploring the legacy of state violence. His films include *Armagh Stories: Voices from the Gaol* (2015) and his publications include *Recording memories from political conflict: A filmmaker's journey* (Intellect, 2010). Mclaughlin and Wills are Co-producers and Co-directors of the film *It Stays With You: Use of Force by UN Peacekeepers in Haiti.*

Alice Neeson is a Research Fellow at the International Communities Organisation. She is particularly interested in community-based action research, social impact, and participatory and creative methodologies. She has a PhD in Social Anthropology, for which her research focussed on the role of storytelling in processes of conflict transformation.

Jacqueline Shaw is a Research Fellow in the Participation, Inclusion and Social Change at the Institute for Development Studies, with expertise using visual methods for participatory action research and community-led change. She started using video for social purposes in 1984, and has since then collaborated with marginalised communities on participatory projects in diverse community, development, and health contexts. She co-authored Participatory Video (Shaw & Robertson, 1997) a definitive methods guidebook, and, as convenor of the *Participate* visual methods programme, worked with partners in India, the Palestinian West Bank, Kenya, and Indonesia to bring the reality of poverty to UN decision makers. Her recent research used participatory visual methods in five countries to explore how to build inclusive and accountable relationships within and across communities, and between marginalised groups' and influential duty bearers. Currently, she is using participatory video to explore resource politics in Kenya for seeing *Conflict at the Margins* (https://seeingconflict.org/about), and contributing to disability inclusion research in seven countries.

Siobhan Warrington is a researcher, writer, and facilitator who works with non-governmental organisations (NGOs), community organisations, universities, and artists. She has an interest in participatory research, representation, and the value of lived experience. She has conducted two multi-country research projects that explore the process of film and photography, and visual representation from the perspective of those featured in NGO communications, She has been commissioned by several international NGOs to produce guidelines and training materials on ethical approaches to image gathering and has produced a set of ethical guidelines for the UK NGO sector. Her article with Dr E. Ademolu titled, *Who Gets to Talk about NGO Images of Global Poverty* was published in Photography and Culture journal Volume 12 (2019).

Siobhán Wills is Professor of Law at the Transitional Justice Institute, Ulster University and primary investigator on the AHRC funded project Community Experience of Conflict in Haiti: Assessing the Emotional Legacy of Civilian Deaths as a result of Intense Use of Force by UN Peacekeepers. Mclaughlin and Wills are Co-producers and Co-directors of the film *It Stays With You: Use of Force by UN Peacekeepers in Haiti.*

SERIES PREFACE

Ron Iphofen (Series Editor)

This book series, *Advances in Research Ethics and Integrity*, grew out of foundational work with a group of Fellows of the UK Academy of Social Sciences (AcSS) who were all concerned to ensure that lessons learned from previous work were built upon and improved in the interests of the production of robust research practices of high quality. Duplication or unnecessary repetitions of earlier research and ignorance of existing work were seen as hindrances to research progress. Individual researchers, research professions and society all suffer in having to pay the costs in time, energy and money of delayed progress and superfluous repetitions. There is little excuse for failure to build on existing knowledge and practice given modern search technologies unless selfish 'domain protectionism' leads researchers to ignore existing work and seek credit for innovations already accomplished. Our concern was to aid well-motivated researchers to quickly discover existing progress made in ethical research in terms of topic, method and/or discipline and to move on with their own work more productively and to discover the best, most effective means to disseminate their own findings so that other researchers could, in turn, contribute to research progress.

It is true that there is a plethora of ethics codes and guidelines with researchers left to themselves to judge those more appropriate to their proposed activity. The same questions are repeatedly asked on discussion forums about how to proceed when similar longstanding problems in the field are being confronted afresh by novice researchers. Researchers and members of ethics review boards alike are faced with selecting the most appropriate codes or guidelines for their current purpose, eliding differences and similarities in a labyrinth of uncertainty. It is no wonder that novice researchers can despair in their search for guidance and experienced researchers may be tempted by the 'checklist mentality' that appears to characterise a meeting of formalised ethics requirements and permit their conscience-free pursuit of a cherished programme of research.

If risks of harm to the public and to researchers are to be kept to a minimum and if professional standards in the conduct of scientific research are to be maintained, the more that fundamental understandings of ethical behaviour in research are shared the better. If progress is made in one sphere everyone gains from it being generally acknowledged and understood. If foundational work is conducted everyone gains from being able to build on and develop further that work.

Nor can it be assumed that formal ethics review committees are able to resolve the dilemmas or meet the challenges involved. Enough has been written about such review bodies to make their limitations clear. Crucially, they cannot follow researchers into the field to monitor their every action; they cannot anticipate all of the emergent ethical dilemmas nor, even, follow through to the publication of findings. There is no adequate penalty for neglect through incompetence, nor worse, for conscious omissions of evidence. We have to rely upon the virtues of the individual researcher alongside the skills of journal reviewers and funding agency evaluators. We need constantly to monitor scientific integrity at the corporate and at the individual level. These are issues of quality as well as morality.

Within the research ethics field new problems, issues and concerns and new ways of collecting data continue to emerge regularly. This should not be surprising as social, economic and technological change necessitate constant re-evaluation of research conduct. Standard approaches to research ethics such as valid informed consent, inclusion/exclusion criteria, vulnerable subjects and covert studies need to be reconsidered as developing social contexts and methodological innovation, interdisciplinary research and economic pressures pose new challenges to convention. Innovations in technology and method challenge our understanding of 'the public' and 'the private'. Researchers need to think even more clearly about the balance of harm and benefit to their subjects, to themselves and to society. This series proposes to address such new and continuing challenges for both funders, research managers, research ethics committees and researchers in the field as they emerge. The concerns and interests are global and well recognised by researchers and commissioners alike around the world but with varying commitments at both the procedural and the practical levels. This series is designed to suggest realistic solutions to these challenges – this practical angle is the *unique selling proposition* (USP) for the series. Each volume will raise and address the key issues in the debates, but also strive to suggest ways forward that maintain the key ethical concerns of respect for human rights and dignity, while sustaining pragmatic guidance for future research developments. A series such as this aims to offer practical help and guidance in actual research engagements as well as meeting the often varied and challenging demands of research ethics review. The approach will not be one of abstract moral philosophy; instead it will seek to help researchers think through the potential harms and benefits of their work in the proposal stage and assist their reflection of the big ethical moments that they face in the field often when there may be no one to advise them in terms of their societal impact and acceptance.

While the research community can be highly imaginative both in the fields of study and methodological innovation, the structures of management and funding, and the pressure to publish to fulfil league table quotas can pressure researchers into errors of judgment that have personal and professional consequences. The series aims to adopt an approach that promotes good practice and sets principles, values and standards that serve as models to aid successful research outcomes. There is clear international appeal as commissioners and researchers alike share a vested interest in the global promotion of professional virtues that lead to the public acceptability of good research. In an increasingly global world in research terms, there is little point in applying too localised a morality, nor one that implies a solely Western hegemony of values. If standards 'matter', it seems evident that they should 'matter' to and for all. Only then can the growth of interdisciplinary and multi-national projects be accomplished effectively and with a shared concern for potential harms and benefits. While a diversity of experience and local interests is acknowledged, there are existing, proven models of good practice which can help research practitioners in emergent nations build their policies and processes to suit their own circumstances. We need to see that consensus positions effectively guide the work of scientists across the globe and secure minimal participant harm and maximum societal benefit – and, additionally, that instances of fraudulence, corruption and dishonesty in science decrease as a consequence.

Perhaps some forms of truly independent formal ethics scrutiny can help maintain the integrity of research professions in an era of enhanced concerns over data security, privacy and human rights legislation. But it is essential to guard against rigid conformity to what can become administrative procedures. The consistency we seek to assist researchers in understanding what constitutes 'proper behaviour' does not imply uniformity. Having principles does not lead inexorably to an adherence to principlism. Indeed, sincerely held principles can be in conflict in differing contexts. No one practice is necessarily the best approach in all circumstances. But if researchers are aware of the range of possible ways in which their work can be accomplished ethically and with integrity, they can be free to apply the approach that works or is necessary in their setting. Guides to 'good' ways of doing things should not be taken as the 'only' way of proceeding. A rigidity in outlook does no favours to methodological innovation, nor to the research subjects or participants that they are supposed to protect. If there were to be any principles that should be rigidly adhered to they should include flexibility, open-mindedness, the recognition of the range of challenging situations to be met in the field – principles that in essence amount to a sense of proportionality. And these principles should apply equally to researchers and ethics reviewers alike. To accomplish that requires ethics reviewers to think afresh about each new research proposal, to detach from pre-formed opinions and prejudices, while still learning from and applying the lessons of the past. Principles such as these must also apply to funding and commissioning agencies, to research institutions and to professional associations and their learned societies. Our integrity as researchers demands that we recognise that the rights of our funders and research participants and/or subjects are to be valued alongside our cherished research goals and seek to embody such principles in the research process from the outset. This series will strive to seek just how that might be accomplished in the best interests of all.

ACKNOWLEDGEMENTS

Thank you very much Ron Iphofen for recruiting me for this extremely rewarding task. I am very grateful for his unwavering confidence in me as we worked towards the completion of this volume. Additionally, this volume would not have happened without the support of Jennie Carlsten, who contributed her expertise in film scholarship and who assisted me with the unruly task of peer editing. She served as a very valuable mentor in this process, and I am very grateful for her time, support, and guidance.

INTRODUCTION: ETHICS AND INTEGRITY IN VISUAL RESEARCH METHODS

Savannah Dodd

ABSTRACT

This chapter serves as an introduction to the key themes found within the volume Ethics and Integrity in Visual Research Methods, *and provides a rationale for the volume's focus on photography and film media. Drawing from other literature, the author discusses the significance of indexicality and visual language when working with photography and film in research contexts, and describes how these considerations set photography and film apart from other forms of visual data. The chapter concludes by outlining the format of the volume, which divides the nine chapters into three key areas of exploration: Voice and Agency, Power and Inequality, and Context and Representation.*

Keywords: Visual data; indexicality; visual language; research methods; photography and film; ethics and integrity

Contemporary society is overflowing with images. Images dominate our public spaces. Full-colour billboard advertisements line our highways. Our free time is spent in cinemas, on video-streaming websites, and in social media networks that use images as a central mode of communication. It is no wonder, then, that there has been a growing desire to harness the power of visual media in research over the past decade.

Visual research methods include, but are not by any means limited to: the study of 'found' and archival visual data; the creation of visual data by research participants (including 'photovoice'); participatory creation of visual data that

Ethics and Integrity in Visual Research Methods:
Advances in Research Ethics and Integrity, Volume 5, 1–8

ISSN: 2398-6018/doi:10.1108/S2398-601820200000005002

involves co-production between researchers and participants; the use of visual data in interviews ('visual elicitation') (Pauwels, 2015); and the representation of research through visual data produced by the researcher. Visual methods have enabled researchers to access different kinds of information (Harper, 2002), to record 'thick description' (Kharel, 2015), and to break down power relationships between the researcher and the research participant (Niskac, 2011). While this increase in the use of visual methods represents exciting opportunities for conducting and presenting research in new and innovative ways, such methods are not without their challenges to ethics and integrity in research.

This volume aims to unpack the multiple considerations for ethics and integrity that accompany different methods of visual data generation and analysis. Contributing authors share their experiences of working with visual methods, and make recommendations for best practice. The authors go beyond the idea of 'one-size-fits-all' guidelines, and instead offer advice for contextualised approaches for addressing the unique challenges presented by visual research methods. More than anything, this volume is intended to be practical, rather than theoretical. Although many of the concepts put forward will be couched in theory, the key messages aim to be concrete and applicable for anyone embarking on a visual methods research project.

ZOOMING IN ON PHOTOGRAPHY AND FILM

Although visual methods can include a wide variety of artistic media (such as painting, drawing, and sculpture) in the generation and analysis of visual data, this volume focusses particularly on photography and film because of the unique relationship these media have with the idea of truth in representation. Unlike drawings or paintings, photography and film are indexical, meaning that they have a direct relationship with the 'real'.

The concept of photographic indexicality, a term given to us by Peirce's (1998) study of semiotics, was based on the technological process of photographic image creation through an analogue camera, whereby the image was literally created by the light that bounced off the subject and onto a photosensitive surface. Early versions of the camera first emerged in the early 1800s, around the same time as the rise of positivism, the theory that the truth is directly observable. Significantly, this further coincided with the colonial project and resulted in the application of photography as an instrument of colonialism, 'beginning with the photographic application to anthropometry and intimately connected to the acts of appropriation and objectification' (Jackson & Dodd, 2019). Photography's indexical nature, combined with positivist theory, has meant that photography has traditionally been trusted to represent the truth. In the colonial context, trust in photographs that portrayed others as inferior validated the colonial project in a way that paintings or drawings could not (Jackson & Dodd, 2019; Mabry, 2014).

Today, people may be less inclined to put their trust in photographs. The emergence of post-positivism by the mid-twentieth century has made us aware of that our ability to observe the world around us is fallible, and, consequently, that there is a lot of subjectivity at play in taking a photograph. While a photograph cannot be understood to represent the 'truth' of an event, it can be understood to represent 'reality'. It was Kracauer (1993) who made this significant distinction that separated indexicality from truth when he wrote that a photographic image cannot give us any greater understanding of a historical event beyond the reality of the 'spatial configuration of a moment' (p. 431).

A photograph might be an accurate reflection of the literal placement of objects within the frame, yet the information that a photograph relays is complicated by the various conscious and subconscious decisions the photographer makes during the process of taking a photograph. By framing an image through a camera's lens, the photographer chooses aspects of reality to capture and aspects to omit. Although the 'reality' of the photographs will be accurate, there is a risk of over- or under-representing certain aspects, thereby misrepresenting the truth of an event.

Brothers writes that indexicality, or perhaps we can understand the use of indexicality here as the 'spatial configuration of a moment', is one way in which a photograph conveys meaning. The second way that a photograph conveys meaning is through iconicity, or the codes and conventions of photographic representation,

> drawn from the fine arts and advertising, from fashion and etiquette, from stereotypes of sex, class, age and race, to name a few, all lifted omnivorously from the culture in which the image is immersed. (Brothers, 1997, p. 18)

She writes that accessing the meaning of a photograph comes from understanding what is literally depicted in the photograph and the iconic functions of a photograph within the context of its production (Brothers, 1997, p. 22).

The first kind of icon that is worth considering here is the icon of indexicality. Tietjen (2018) argues that, although most photographs are now taken using digital cameras, photographs are still imbued with a special relationship to reality because photographs have become icons of index. Although modern-day digital photographic technologies no longer have the same relationship to the real through their production process and can be more easily manipulated, indexicality 'is not a quality intrinsic to analog images and absent from digital ones' (Tietjen, 2018, p. 377). He ties indexicality instead to our recognition of the iconicity of a photograph: we understand that 'that' is what a photograph is and that a photograph is indexical.

However, indexicality is not the only kind of icon at play in a photograph. In addition to conscious and subconscious decisions about what to include or exclude when framing a photograph, the photographer makes a host of other decisions that shape the representation of an event, from determining the perspective from which to take the shot to deciding how to edit a photograph after it has been taken. These decisions, often considered as simply aesthetic preferences, contribute to the visual language that the photographer uses to communicate meaning.

When anyone, researcher or research participant, takes a photograph they are consciously or subconsciously drawing on their knowledge of visual culture, or the shared system of visual representation and communication in a society (Sturken & Cartwright, 2001, p. 3). Yet, as Rose (2014) writes,

> Participants are asked to draw maps and take photographs and make films as if they had never opened an A to Z, seen a family snap or been to the movies. (p. 11)

A lot of visual data research fails to account for the relationship between visual culture and the visual research methods they are implementing (Rose, 2014, p. 5). Such research is largely concerned with the content of photographs, 'the visible', and gives little consideration to visual language, 'the visual', or the iconicity invoked with those photographs (Rose, 2014, p. 10).[1]

Whether generating, analysing, or co-producing visual data, researchers would be wise to give greater credence to visual language, and to critically consider the etymology of visual language, because even the seemingly innocuous aesthetic choices impact the meaning conveyed in a photograph. For example, choosing to photograph a person from above creates a top-down perspective that might convey a sense of inferiority of the individual pictured, or of superiority of the viewer or photographer. A preference for a top-down perspective when photographing Ugandan children could be understood as an inherited visual trope from the colonial era that used these aesthetics to reinforce existing power relations. This would be a useful frame through which to analyse such a photograph, and it would be an important consideration when choosing a perspective from which to take a photograph.

Since photographs and film, in contrast to paintings or drawings, are considered to be indexical and, therefore, representative of reality, there is added responsibility to ensure that the meanings conveyed through the visual language invoked are accurate and appropriate. While this discussion of indexicality and visual language has focussed on the medium of photography, these concepts are also very relevant to film. In fact, film presents additional considerations that are unique to the moving image, including sequencing and the accompaniment of sound (Davies, 2011; Dymek, 2013; Nichols, 2016). These kinds of structural components involved in image-making impact on the way that meaning that is made through still and moving images. A lack of consideration for these components can present threats to both ethics and integrity in research, especially when using visual data produced by researchers or for the purpose of representing research outcomes. Although indexicality and visual language are not primary concerns of many of the contributing authors in this volume, this discussion is critical to justify focussing this volume on photography and film methods.

ETHICS AND INTEGRITY OF PHOTOGRAPHY AND FILM IN RESEARCH

Having your photograph taken can feel very awkward. It can be uncomfortable to be at the end of a camera lens, reflecting your face in that dark mirror, knowing that you are being looked at, and that you will be represented in a photograph for others to look at. It is liable to feel even more invasive if you are photographed in your own home or during an illness or when you are at your most vulnerable.

Therefore, the decision to use a photographic or filmic method in a research project must be carefully weighed. Before deciding to incorporate such a visual element into a research project, it is important to determine whether taking photographs or recording film will provide a piece of research with something that could not otherwise be accessed. What are the theoretical grounds for including photography or film in the research design? How will photography or film uniquely contribute to answering the research question? Is photography or film the best visual method for the context? Although photography and film can be extremely useful tools, it is important to remember that these visual methods might not always be appropriate for a research project due to context, relevance, or other important factors.

Once it has been established that there is a valuable and legitimate reason for potentially exposing research participants to the kind of vulnerability that comes with being photographed or filmed, consideration must go into deciding how to approach such a project. There is an undeniable power dynamic between the person photographing and the person being photographed, and this power dynamic must be well managed when collecting visual data. When approached with care, photography and film elements to research can be rewarding for all parties involved, but when handled without consideration for the experience of the participants, it can be extremely damaging.

The following chapters will discuss in detail the in's and out's of approaching, designing, and executing research using visual data in a variety of ways. Before moving into the body of this work, it is important to introduce some key terms that will be significant throughout the volume.

The researchers writing for this volume come from a variety of backgrounds and practice research in a variety of ways. Therefore, the term 'research' in this volume casts a wide net to include basic, applied, action, and practice-led research (Smith & Dean, 2009). Ethics and integrity are concepts that are integral to all types of research, where ethics is the 'balancing harms and benefits' in an 'attempt to ensure harm is minimised or avoided and that benefit is maximized' (Iphofen, 2013, p. 13) and integrity is 'a commitment to intellectual honesty and personal responsibility for one's actions and to a range of practices that characterize responsible research conduct' (Committee on Assessing Integrity in Research Environments, 2002, p. 34). The question is: how can we apply them to research with visual data? Even further: how might practices to ensure ethics and integrity change when applied to visual data?

VOLUME OVERVIEW

Although there are seepages and overlaps between these categories, I have divided the chapters into three key areas of exploration: Voice and Agency, Power and Inequality, and Context and Representation. In the first section titled Voice and

Agency, authors contribute to discussions about the ownership of images, the silencing quality of anonymisation, and the danger of paternalism in risk management.

In Chapter 1, Aura Lounasmaa, Cigdem Esin, and Crispin Hughes reflect on the challenges they faced while facilitating visual story-telling workshops in the Calais Jungle refugee camp in 2016. They illustrate the complexity inherent in balancing a decolonising practice that values authorship and ownership, with the practicalities of managing the risk of jeopardising the participants' future resettlement in the European Union. The authors explain how their multimodal approach created opportunities for researchers and participants to interrogate the asymmetrical power relationships that characterise life both within and outside of the Jungle. They conclude with the observation that listening is key to ethical and decolonial practice.

In Chapter 2, Angela Stephanie Mazzetti outlines the limitations to protecting the identities of research participants when using photographic methods, and explains the potential risk for participants when their identities not protected, especially when depicted as engaged in illicit activity. Drawing from her experience conducting research in the post-conflict context of Northern Ireland, she proposes that participant-produced drawing can serve as an effective alternative to photography in ethnographic research. Drawing enables the participant to deliberately select elements for inclusion in their image, without the risk of including identifying information that may be accidently recorded in a photograph.

In Chapter 3, Siobhan Warrington concludes the section on Voice and Agency by describing research she conducted that aimed to highlight the voices of individuals who are pictured in NGO communications' materials. Her chapter traces the ethical decisions made during the course of research for Save the Children's 2017 report *The People in the Pictures: Vital perspectives on Save the Children's image making.* Warrington explains how she used photo elicitation methods in interviews and focus groups, highlighting the ethical risks of the method and the ways in which she worked to mitigate those risks. Reflecting on her research practice, she concludes by offering key points for consideration to researchers who are planning to use photo elicitation methods. She also draws on her research findings to offer recommendations for responsible image-making practice.

In the second section titled Power and Inequality, contributing authors explore methods of challenging unequal power relationships in visual research processes, and of restoring control to the individuals pictured in visual research outputs through long-term engagement with a community.

In Chapter 4, Robert Godden identifies informed consent as being at the intersection of the rights and the obligations that non-profit organisations have towards the individuals pictured in their communications materials. Godden begins by defining informed consent, and by identifying the risks and challenges, especially with regard to the extra layers of consideration needed when working with children. He then engages in a dialogue with freelance photojournalist Smita Sharma to identify practical solutions for the successful implementation of informed consent in image gathering.

In Chapter 5, Cahal McLaughlin and Siobhán Wills use their experience of producing a film about the violence inflicted by UN Peacekeeping Troops in Haiti as a case study for understanding ethics and integrity in research-based filmmaking. The authors explain the challenges they encountered both at the production stage and at the dissemination stage of the resulting film *It Stays With You: Use of Force by UN Peacekeepers in Haiti.* In particular, they discuss the challenges of applying a participatory approach to filmmaking in the face of barriers like language, distance, and resource inequality.

In Chapter 6, Jacqueline Shaw advocates for long-term, as opposed to short-term or one-off, projects when using visual methods as a more ethical mode of engagement, especially among marginalised communities. In particular, she advocates for long-term, progressive engagement in participatory video projects involving the creation of video content for both internal and external audiences. Creating content for different audiences enables participants to make more informed decisions about consent and to further develop their ideas about the given topic through multiple iterations.

In the third section titled Context and Representation, authors discuss practical considerations for context in the production and dissemination of visual research, and the ways in which context, or a lack of contextual information, can change the meaning of visual data.

In Chapter 7, Ciara Chambers analyses the different ways in which practitioners have recontextualised visual data by using archival material in the production of new films. She describes how these filmmakers have approached the archive, how they have negotiated relationships with relevant stakeholders, and how they have responded to the ethical challenges that have arisen as a result of reuse. She concludes by applying this knowledge to the research process, making recommendations for an ethical approach to the reuse of archival material in research contexts.

In Chapter 8, Susan M. Brigham and Mohamed Kharbach draw on their experience facilitating a participatory photography project among youth with refugee experience in Canada. They outline the ethical considerations that emerge stage-by-stage in research processes using participatory photography methods, drawing attention to the special considerations needed when conducting visual research with youth populations and in transcultural contexts. In particular, Brigham and Kharbach assert the need to situate ethics within the socio-cultural contexts of the production and the reception of research images.

In Chapter 9, Alice Neeson explains that the objectives of action-focussed research often lay in the processes of visual data production, rather than in the analysis of the resulting visual data. However, she highlights the potential ethical implications of disseminating visual data without including analysis. Neeson explains that the presentation of visual data without context or caption leaves them open to the interpretations of viewers who may not have access to important background information. This is further complicated by the online circulation of visual data, which allows viewers to appropriate and create their own pathways to visual data.

In Chapter 10, Jennie Carlsten describes the challenges of image-based research, in particular research focussed on the study of images created by others, within institutional contexts whose frameworks are often not relevant for or applicable to work with visual data. She draws from ethical philosophy to situate the institutional approach to ethics within a wider debate, and highlights alternative models including 'situated ethics'. Carlsten calls for more flexible approaches to research ethics, and for a shift in research practice that is more collective, reflective, and empathetic.

NOTE

1. A notable exception is research that uses archival visual data, for example, the works of Brothers (1997), Banks and Vokes (2010), and Campt (2012).

REFERENCES

Banks, M., & Vokes, R. (2010). Introduction: Anthropology, photography and the archive. *History and Anthropology*, *21*(4), 337–349.

Brothers, C. (1997). *War and photography: A cultural history*. London: Routledge.

Campt, T. (2012). *Image matters: Archive, photography, and the African diaspora in Europe*. Durham: Duke University Press.

Committee on Assessing *Integrity in Research Environments. (2002). Integrity in Scientific research: Creating an environment that promotes responsible conduct*. Washington, D.C.: National Academies Press (US). Retrieved from https://www.ncbi.nlm.nih.gov/books/NBK208712/pdf/Bookshelf_NBK208712.pdf

Davies, D. (2011). Digital technology, indexicality, and cinema. Rivista di estetica, *46*. [online] Retrieved from http://journals.openedition.org/estetica/1655

Dymek, A. (2013). Perception, dreams, films: Iconicity and indexicality in Peirce's theory of perception. *Recherches Sémiotiques*, *33*(1-2-3), 39–61.

Harper, D. (2002). Talking about pictures: A case for photo elicitation. *Visual Studies*, *17*(1), 13–26.

Iphofen, R. (2013). Research Ethics in Ethnography/Anthropology. *European Commission's Directorate-General for Research and Innovation*. [online] Retrieved from https://ec.europa.eu/research/participants/data/ref/h2020/other/hi/ethics-guide-ethnog-anthrop_en.pdf

Jackson, A., & Dodd, S. (2019). The Impact of the White, Male Gaze. *Trigger*, 1(Impact), 43–49.

Kharel, D. (2015). Visual ethnography, thick description and cultural representation. *Dhaulagiri Journal of Sociology and Anthropology*, *9*, 147–160.

Kracauer, S. (1993). Photography. (T. Levin, Trans.) *Critical Inquiry*, 19(3), 421–436.

Mabry, H. (2014). Photography, Colonialism and Racism. International Affairs Review. [online] Retrieved from https://www.usfca.edu/sites/default/files/arts_and_sciences/international_studies/photography_colonialism_and_racism_-_university_of_san_francisco_usf.pdf

Nichols, B. (2016). *Speaking truths with film: Evidence, ethics, politics in documentary*. Oakland: University of California Press.

Niskac, B. (2011). Some thoughts on ethnographic fieldwork and photography. *Studia Ethnologica Croatica*, *23*(1), 125–148.

Pauwels, L. (2015). *Reframing visual social science: Towards a more visual sociology and anthropology*. Cambridge: Cambridge University Press.

Peirce, C. S. (1998). In N. Houser, C. Kloesel, & the Peirce Edition Project (Eds.), *The essential Peirce* (Vol. 2). Bloomington: Indiana University Press.

Rose, G. (2014). On the relation between "visual research methods" and contemporary visual culture. *Sociological Review*, *62*(1), 24–46.

Smith, H., & Dean, R. (2009). *Practice-led research, research-led practice in the creative arts*. Edinburgh: Edinburgh University Press.

Sturken, M., & Cartwright, L. (2001). *Practices of looking: An introduction to visual culture*. Oxford: Oxford University Press.

Tietjen, F. (2018). Post-Post-Photography. In M. Neumüller (Ed.), *The Routledge companion to photography and visual culture* (pp. 376–378). New York, NY: Routledge.

PART I

VOICE AND AGENCY

CHAPTER 1

DECOLONISATION, REPRESENTATION, AND ETHICS IN VISUAL LIFE STORIES FROM THE JUNGLE

Aura Lounasmaa, Cigdem Esin and Crispin Hughes

ABSTRACT

This chapter discusses ethics in participatory photography with focus on refugee participants and informal refugee camp setting. The chapter draws on ethics in participatory photography projects elsewhere and especially the experiences of photographers who work with these methods. The context here is the Calais Jungle camp, where the authors worked with a group of participants, who were residents of the camp, over several months to encourage photographing and documenting life in the camp and beyond, and to work on life stories that can be drawn from and inspired by these photos. The project, and hence the ethics in our work, were framed by the experiences of the refugee participants, and so at all times the authors needed to navigate temporality, violence, state oppression, lack of resources, human rights violations, language barriers, religious and cultural differences, national and supranational immigration policies, shame, and more. This chapter discusses how the authors navigated these ethical issues, the limitations of the approaches and solutions they found, and the lessons they learned, which can be applied to research using participatory visual methods with refugees.

Keywords: Ethics; refugees; photography; narrative methods; participative research; jungle

Ethics and Integrity in Visual Research Methods:
Advances in Research Ethics and Integrity, Volume 5, 11–28

ISSN: 2398-6018/doi:10.1108/S2398-601820200000005004

In 2016, the University of East London ran visual storytelling workshops in the Calais Jungle refugee camp. The visual life story workshops were part of a larger education project in the camp, through which we encouraged refugees to tell, share, and make sense of their life stories using multimodal narratives. The aim was to facilitate the creation of visual, verbal, written, and processual narratives, respecting their past, their history of displacement, and present conditions, in contrast to the misrepresentation in the media. The edited images and stories were exhibited in physical and online spaces, and many were collated in a book, co-authored by the residents, *Voices from the Jungle* (Calais Writers, 2017). Crispin Hughes worked on the project as a photographer, Aura Lounasmaa coordinated access to the camp and taught on the other educational projects there, and Cigdem Esin helped design the multimodal perspective and methodology for the project, together with Corinne Squire, who started the project (Esin & Squire, 2013). Various others helped run the projects, which took place over a year in the Calais Jungle, before it was dismantled in the autumn of 2016, and we are grateful for all of them as well as the participants for their contributions.

The project was not framed as research, as we did not wish to become ones who 'take stories' from the residents of the Jungle – already in the public eye, yet unable to set the limits of the representations that are made of them. Instead, we wished to come to the camp with an open mind, offering to those we met there our skills and knowledge through a university course, without any expectations or pressure to produce outputs from the project. Participation was entirely voluntary, and while some participants wanted to share their stories at the end of the project publicly, they did so as authors of their own work rather than as research participants. What we learned through the processes of setting up and running these workshops is applicable to research with refugee communities. It has also greatly challenged our assumptions about what research is, and how we might wish to go about it in the future. The team had worked in various participatory visual projects prior to the Jungle, and the political and ethical boundaries here posed a series of new challenges.

We will begin by explaining the political and psychosocial context of the Calais Jungle at the time of the projects. Due to its geopolitical location, developments in European migration policies, and the unique set-up of the camp, a very particular set of issues arose from working with the residents. Our approach is framed by narrative, and particularly multimodal narrative research and visual storytelling. This approach recognises the interrelations that constitute the narratives, including participants' individual and familial stories, their experiences as refugees, as members of transnational communities, and the relations within and across the projects. Using multimodal narratives, a combination of written and visual narratives, within the context of these projects enabled us to facilitate opening up relational spaces for both participants and us. These spaces are shaped by multiple power relations in the field, understood within a Foucauldian framework to include productive consequences, where complex and multilayered interactions emerged. We will explore what these terms mean in the context of our project. Other theoretical issues important to consider here include decolonising methods and pedagogical practices. We also discuss existing research into the ethics of participatory research with refugees, especially in refugee camp contexts.

The following section of this chapter considers four main problems we faced with the project:

(1) The Dublin Accord (Regulation 604/2013 of the European Union and the European Commission) requires refugees to seek asylum in the first safe country they arrive in. The fear of being returned to France, if they managed to reach the UK, meant participants could not be safely identified as residents in the Jungle. This meant that recognisable self-portraits and authorship of the images and stories had to be erased.
(2) Even as participants had fled war, persecution, torture, and imprisonment, the Jungle had gained such an infamous reputation that many participants felt compelled to hide the fact they had ended up in this dire situation from their families: instead of using their photos to tell stories of the Jungle, they used them to represent an aspired new life in France. This made it difficult to negotiate common aims for the storytelling, and created issues regarding anonymity and representation.
(3) Ownership of photos, stories, and cameras remained with the participants throughout, but as each moved on and the Jungle was dismantled, staying in touch with some became impossible. Hence not all voices could be heard/represented. The ethical requirement for continuous consent carries to the present, and the future, and may have implications for both us and our participants that we are only beginning to see now.
(4) With all efforts to make the project participatory and decolonise the educational practices, issues of language competency, translation, and expert knowledge were present in the workshops. While the aim was to move away from the negative media representations, life stories were still shaped by those who elicited them.

The chapter will consider these questions with reference to existing research and the lessons we drew from our own experiences and exchanges with our participants to this date. We draw here from our reflections of the processes and experiences running the programmes, and from publicly available accounts and images, published online or in a book written by Jungle residents and edited by our team, *Voices from the Jungle* (Calais Writers, 2017). Lastly, we will share our reflections on the unresolved ethical dilemmas and issues to which this project alerted us, and make some suggestions for future participatory photography projects with refugee participants.

CALAIS JUNGLE

In the context of the Jungle refugee camp, the legal and political frameworks that limit refugees' full participation in society also put ethical and practical limits on any type of project that was run with the refugee residents of the camp. In this chapter, we elaborate on reflections on the collaborative work we did with refugees, and the methodological choices should be discussed in relation to ethical decisions and positions we took while working with participants who have limited access to citizenship rights and information.

As the conflicts in Syria, Yemen, Iraq, and various states in North Africa continue over years, a record number of people are displaced either internally, or across continents. This raises a great number of immediate needs to be met, which are often prioritised above the long-term needs to continue their lives in the settlement countries where they reside temporarily or permanently. Given the acute needs for survival in the camps and during their settlement period, social and educational needs may be viewed as secondary. The importance of education is emphasised for future reconstruction and stability for both conflict regions and post-migration countries (Butler, 2016). Yet, many refugees face barriers of recognition in terms of social and professional qualifications even after their settlement (Houghton & Morrice, 2008). The material conditions in refugee camps and the urge to survive often push refugees to adjust to power relations that shape their everyday lives, leaving their connection to their past and future aside.

Although Calais has been a place where refugees and migrants have gathered to try and cross the English Channel and reach the UK for many years, the recently elevated conflicts and unrest in the Middle East and parts of Africa have severely increased the number of refugees in Europe. In Calais too the number of people regularly awaiting passage across rose from some hundreds in early 2015 to as many as 10,000 in 2016.[1]

We wish to point out here that we are using the term 'refugee' throughout this chapter to refer to participants of the photography workshops we ran in the Calais Jungle in 2016, all of whom were forced migrants and residents of the unofficial camp at some point between 2015 and 2016. The legal term refugee usually refers to those who have already received refugee status and legal protection from a state. This was not the case for most of our participants and students at the time of the photography workshops. Yet, the circumstances of the camp, the migration journeys, and the legal constraints of the participants framed the project in such a way as to emphasise the experiences central to their refuge-seeking in the narratives that were produced. The negative press coverage (Finnish Institute in London and the Finnish Cultural Institute for the Benelux, 2016), especially in the UK, led *Al Jazeera* among others to stop using the term migrant to describe those seeking asylum in Europe (Malone, 2015). The term asylum seeker would indicate that all participants had already sought protection, which also wasn't the case at the time of the project, as some were still planning to seek asylum outside of France.

It was this possible future mobility which framed one of the major ethical and safeguarding concerns we had during the photography and writing projects. The Dublin regulation (Regulation 604/2013 of the European Union and the European Commission) stipulates that anyone seeking asylum in the European economic area must do so in the first EU member state they enter. If proof exists that a person has already passed through another member state, they can be returned to that state without the asylum case being considered. Often this proof is fingerprints taken by border agents, but other official documents or even photographs could become the required proof in some instances. Safeguarding the anonymity of our participants was hence a priority at all times.

The so-called 'refugee crisis' of 2015 and 2016 was widely reported in European media. Calais Jungle became a favourite destination of UK press due to its proximity, easy access, and the sensationalist headlines it promised due to poor living conditions and clashes between camp residents and lorry drivers and police and civilians. Residents quickly learned to mistrust the press and others aiming to document lives in the camp. The Jungle also attracted research and other projects focussing on lives of refugees in temporary settlements, which have been proliferated also elsewhere by the rapid increase in populations crossing borders to reach Europe. Documenting refugees' lives and practices has been part of several endeavours which aim to understand conditions of refugees and/or support them in their journeys (see, e.g. Sanyal, 2017). All these projects have been conducted in complex and sometimes difficult and dangerous conditions. Hugman, Pittaway, and Bartolomei (2011) underline the possibility of putting the refugee communities at risk of harmful research practices under such complex conditions even though such outcomes might be unintended. While following the baseline of 'do no harm' is recognised as a path to follow in work with refugees or other marginalised groups, it is essential to follow a carefully structured ethical-reflexive procedure in order to prevent any outcomes which could make the lives of these groups even more difficult.

MULTIMODAL NARRATIVES AND VISUAL STORYTELLING

Storytelling has the potential to enable refugee participants to make sense of their lives, to connect their past, present, and future while they are in the middle of a crisis. Telling life stories could also provide an alternative to the issue of representation where language and cultural difference remain as a barrier in a historical context in which refugees face hostility, marginalisation, and discrimination. We used narrative strategies while working with participants, approaching life narratives in any form (spoken, written, and visual) as a tool for the storyteller in the process of self-making and world-making (Bruner, 2003), in response to a network of power relations. For other researchers who work with narratives, it is important to constitute a research process which allows them to listen to the stories, and to interact with the participants in order to explore their meanings and storytelling.

It may be proposed that the creative methods and multimodal narratives expanded the space of the interactions and thus enabled the multiple participants of the projects, including ourselves, to consider the broader context of the asymmetrical power relations in which we lived and worked. Our multimodal approach was shaped by Esin's (2017) previous work with young Muslim women in the UK, where she used the approach to open up creative spaces and narrative construction of the self with these participants. Here young Muslim women came together with the visual artist Chila Burman to create self-portraits using collage. Participants were then interviewed about their collages, and the images themselves included in the analysis as important pieces of the told stories. It is difficult to claim that a power balance shaped the interaction within the space of the projects. It is much more complex. But the creative spaces we managed to open up together with participants of the projects functioned as spaces in which participants were able to express their recognition of otherness but also to challenge the constitution of ideas about 'us and them' within the broader relations of power.

Moreover, conversations within these spaces led to further questions around the availability of resources that the participants could use to contest the asymmetrical power relations in which they lived, their access to social and political resources to challenge broader inequalities. We considered the space of the projects as a space to hear the 'voice of the refugee/forced migrant', which can challenge the competing voices that come from more socially powerful exogenous agents which may often discount or minimise the refugee experience (Benezer & Zetter, 2014). In this way, 'the approach [creative visual method] is optimistic and trusting about people's ability to generate interesting theories and observations themselves' (Gauntlett & Holzwarth, 2006, p. 82). Of course, hierarchies of power and access operated on different levels at all times, and we were aware these intersectionalities and what they meant for our own safety and access to the camp at different times, the stories we would be able to hear and people who would speak to us, and in turn how our participants were differently affected by all this. Emphasis on narrative and visual methods, however, highlighted the participants' agency in choosing what to reveal and to whom.

Visual storytelling reflects social, cultural, and political processes, individuals' meanings attributed to their life circumstances and experiences. The visual part of this storytelling practice has the potential to connect (although temporarily) times and spaces in narratives that are otherwise distant (Gastaldo, Magalhães, Carrasco, & Davy, 2012). Researchers who work with refugees have been increasingly using photography as a participatory method. For example, Gomez and Vannini (2015, 2017) used *Fotohistorias* as a method to examine and map out migrants' perception of home and sense of belonging in Americas in photographs about their experiences of migration. Participants were then invited to take part in interviews focussing on the photos they selected themselves. Gomez and Vannini (2017) argue the three main offers that participatory photography could make as a method while working with migrants: (1) taking and selecting photographs encourage participants to take more control over the information they share through the photographs; (2) photographs can widen the space of research conversation between participants and researchers; and (3) producing and discussing photographs could enable participants to externalise their complex emotions, particularly in contexts where they find it difficult to speak up for themselves due to power imbalances.

Barthes (1981 cited in Bignante, 2010) points to the polysemic meaning and interpretations that images could offer. The producers of images could convey uncoded messages and meaning to be deciphered, and these meanings would be multiplied each time a viewer reads/interprets those images. The reading/interpretation of those images mediate the social and cultural narratives through which messages or stories are constituted.

DECOLONIALITY

Some of us constantly put the focus on life stories under scrutiny: what could telling life stories offer to participants in the context of the camp, which was shaped by multilayered power relations, including multiple forms of racism? But having known that we, as facilitators, educators, and collaborators, were part of the same network of power relations as the refugee participants of the projects, we constantly reflected on our own positions and engaged in conversations on the ethical dilemmas that we had to deal with throughout the process.

Action research has, from its first introduction in the 1940s, been linked to grand ideas about social change and social justice through emancipatory and collaborative practices (Collier, 1945). In the education sector, self-reflection to improve our own educational practices is often the main focus (Corey, 1953), changing the nature of education and educational institutions (Rudduck, 1988) which in turn can has the potential for wider emancipatory developments and social change, as the role of educators, learners, and institutions are questioned and analysed and learners empowered to take charge of their education (Freire, 1972). This central aim of empowering learners to move towards social change and equality is at the core of the decolonising project.

As in education, in research, and participatory projects, the first step in decolonising the process must be challenging knowledge hierarchies through dialogue. Drawing on Riessman's (2008) arguments on the dialogical construction of narratives, we approach the visual stories/narratives as products of an ongoing process of dialogue within the projects, with the broader social, cultural, and political narratives. This is an approach to acknowledge the interrelated contexts in which visual stories are co-constructed by the photographer/participants, facilitators, and broader audiences. The dialogic approach is significant to make sense of the power relations embedded in the production and consumption of visual stories. It is through those constantly changing relations that stories become strategic tools to negotiate the position of the participants in response to the political context which shape the lives of refugees.

Within the context of the projects, interconnections between the process of construction and re-construction of visual narratives shape the content and message of the images. Visual narratives are constituted by interrelations between individual and cultural geographies (Doloughan, 2006), and as a space for narrative imagination, which involves continuous negotiation and interaction between the self and the other and between personal and collective thinking, not only reaching out to the future but also deeply rooted in the past (Andrews, 2014).

THE RELATIONALITY OF ETHICS: CONCEPTUAL-METHODOLOGICAL FRAMEWORK

The narrative approach that we used in several projects that we ran with the refugee residents of the Calais camp is interlinked with our consideration of narratives as a form of interaction between storytellers and listeners.

The stories are co-constructed in a relational process, which bring together life experiences and cultural and social resources of both storytellers and the audiences. This interactional process not only opens up a space for the emergence of life stories, but also invites both the storytellers and listeners to explore what is already known to them in relation to themselves and others. This interactional process is very much shaped by the personal, social, and cultural contexts in which we live. It is within the same interactional context that our practice, communications, actions, and ethical positions are constituted. It is crucial to employ a reflexive perspective while creating a narrative of our practice. This means remaining alert to our own positioning while working with camp residents. We recognised our relative power as white, UK-based academic and professional women and man who could move safely in the camp during daylight hours but could not stay beyond sunset. Perhaps the most striking difference between us and the participants was the time they spent waiting to get the UK compared to the one hour it took us on the Eurostar to arrive at London St Pancras station.

Our work had to be grounded in ethical conduct, which would respond to the unpredictability of working conditions while we were at the camp. It was possible to give information to the potential participants about the projects but most of the time it was not possible to provide information about the next stage. This was because most participants did not know whether they would still be in the camp the following morning. New residents and participants arrived daily. Ongoing consent and collaborative planning was impossible in these circumstances and for many people we worked with the project served as a passing distraction. Others, with whom we worked for longer, became embedded in the planning of the online exhibitions and the book, *Voices from the Jungle*, and remain collaborators to this date. Our ethical conduct pushed us to recognise the limitations of the work and the dilemmas, to be open to engage in conversation. That includes, at times, questioning our sensitivity to the rights, beliefs, cultural differences, power imbalances in patriarchal systems, etc., as has been highlighted by Etherington (2007).

Starting with the question of 'what constitutes ethical research practice?' Guillemin and Gillam (2004) identify two dimensions of ethics: procedural ethics and 'ethics in practice'. Procedural ethics usually involves the procedure to follow to get approval from a relevant ethics committee whereas ethics in practices refers to ethical issues that arise while doing research. While examining the relationship between them, the authors explore the notion of 'ethically important moments' where a decision has to be made in response to an ethical responsibility. These moments may not necessarily be linked to an ethical dilemma. According to Guillemanin and Gillam (2004, p. 265), there is a clear response that the researchers could choose in some cases. But there is still something ethically important in those decisions. The ethically important moments emerge within the everyday research interactions, when a researcher has to make a decision when a participant discloses a painful part of their life or when it is essential to respect the dignity of project participants who live in adverse conditions. The authors offer reflexivity as a resource to handle ethically important moments. They approach reflexivity as a process that allows us to consider how our social and political locations affect our practice (Guillemanin & Gillam, 2004, pp. 274–275). It is a process of critical scrutiny and interpretation of the researcher, participants, and the research context. This approach to ethically important moments with the inclusion of a reflexive process into practice is useful when considering ethical questions related to those unanticipated situations and concerns that arise and demand immediate attention in the work we do.

For us, that involves relational ethics (Ellis, 2017) through which we seek answers to questions such as how to make good interpersonal decisions concerning our responsibilities towards those in our studies, when we ask about traumatic, personal events that engender strong emotions in respondents. Relational ethics recognises and values mutual respect, dignity, and connectedness between researcher and researched, and between researchers and the communities in which they live and work (Ellis, 2017, p. 4). The workshop, or indeed the research encounter, should thus become a space in which the focus shifts from finding a solution to ethical problems to asking more and more questions that will help us to build up relations with respect.

Considering ethics as relational requires a systematic examination of the position of the researcher/facilitator. Kofoed and Staunæs (2015) offer ethical hesitancy as a strategy for the researchers in the field when they need to deal with distressed, concerned participants in zones of high intensity. Kofoed and Staunæs (2015) describe moments in zones of high intensity when the researchers are required to make an intervention to solve a problem, to do what is good and take rapid action. In those moments, they suggest, ethical hesitancy may be a more useful strategy. They describe hesitancy as 'a momentary suspension of action due to an embodied sense of thoughtfulness and engaged capability of interrupting one's own immediate incentives to response and enact embedded normativities and judgements' (Kofoed & Staunæs, 2015, p. 25). Hesitancy as ethics enables the researcher to consider their long-term contribution to the communities they work with. Instead of contributing to a clear solution, researchers contribute to the development of a solution without rapid intervention.

While equipped with these important approaches to ethics in research, the Jungle and its residents surprised us and caught us off guard more often than we can describe here. While we often wished to pursue photography, storytelling, and writing in a structured format, our participants were more eager to converse over tea. Nayeri (2019) discusses the ethical dilemma we face when expected to accept hospitality from refugees who live in deprived conditions. Like Nayeri, we hesitated. Accepting these invitations created a bond of mutual respect and dignity, where other stories, including the visual ones, could also be shared. Our process of working with people in the camp, and our process of working with images and stories produced there are always guided by these theoretical approaches and more, the most important of which was the approach to our participants as fellow humans, learners, knowers, and tellers, whom we had the privilege to listen to and to hear.

CHALLENGING REPRESENTATIONS

Our aim was never to 'represent' the narratives of the refugees that we work with. The Calais Jungle was already in everyone's consciousness through the press and news coverage it received. We only reached a small number of people in the camp, and any representation would have been limited. It was also impossible to control where such representations would be taken once produced, and how they might be (mis)interpreted. Once narratives are constructed, once we become audience to the stories, once we carry them to different environments as the 'voices of refugees from the Calais camp', 'representation' arises as a matter to consider. Indeed, some of our participants were motivated by this possibility to be differently represented through their own visual and written stories. As Shaheen from Afghanistan stated: '… it was interesting to write, especially because I knew that now, many people would read my story' (Calais Writers, 2017, p. 256). The possibility of representing a different point of view and challenging the harmful media representations of the Calais Jungle and its residents were a strong motivation for many participants throughout the project.

The voice of the refugee heard through their life stories is important to document, and through it we can begin to hear the reality of countries of origin and of refuge. Those voices are critical tools in challenging the stereotypical images of refugees particularly in countries of refuge that are created and sustained by media. The life stories are tools for refugees to build up an active stance showing their resourcefulness and motivation. Refugee narratives are also products of the heterogeneity of refugee experiences. The question of representation then becomes one of the many 'ethically important' moments in the project. Doná (2007) points to the essentialist ways of representing refugee voices through binary categories such as victim/survivor, resilient/vulnerable, deserving/undeserving. The narrative approach that we used was helpful in problematising these essentialist ways in which life stories of refugees from the Calais camp were represented in the media. This approach enabled us take a critical position towards the definition of those binary categories and to respect the complex individuality of each life story. Yet, it is not possible to fully control the destination of stories. When we listened to the stories of refugee participants, we heard details of their past life at home, their painful journeys, their life at the camp, their disappointments and friendships, and their hopes and plans for the future. Being audience to those stories left us with constant worry about who else would listen to their stories. This is not only a concern about the confidentiality of the content, what and what not to reveal while disseminating stories. Our awareness of the emotional investment that refugee storytellers make in their stories keeps us alerted whenever we share stories.

The participants of photography workshops had an intention to represent the camp conditions in their pictures, as a strategy to communicate with the broader world. Yet, they also had difficulties to frame those representations in attempt to challenge the stereotypes and prejudice attached to refugees and their lives. Drew, Duncan, and Sawyer (2010) underline a similar difficulty that the participants in their health research with young people in which they used visual storytelling as a method. The young participants with chronic illnesses found it difficult to photograph the conditions of their everyday life with illness as they did not know how to deal with the stigma attached to the chronic illness they had. Indeed, many of the photographers in the Jungle moved away from photographing the conditions of their life. Many produced images of an imagined life in Europe, where the Jungle did not exist. Many did this to send proof to families back home that they were doing well. By 2017, the Jungle had become so infamous that participants felt they needed to hide the fact they were living there from their families. As Habibi (from Afghanistan) states:

> My parents are in Afghanistan. I didn't tell them, 'I'm living in the Jungle'. When my mom calls, I say, 'Actually yeah, they gave me a very nice house here'. Sometimes they tell me, 'Send me a picture'. I go to a volunteer's house, taking pictures of it, sending them to them. As if I have a really nice house. I cannot tell them the truth, that I'm living the 'Jungle' life. (Calais writers, 2017, p. 116)

In addition to pictures of the nice houses and cars, participants posed for pictures in front of monuments, statues and grand buildings, as if on a holiday touring France. These images were interesting and communicated an important aspect of the imagined and aspiring life stories of the participants, but posed the issue of anonymity and safety of the photographers. How could we anonymise and safeguard the participants while simultaneously respecting their authorship and integrity as photographers?

DECOLONISING PRACTICE, PHOTOGRAPHY, AND ETHICS

Prior to the collaboration between the authors in the Calais Jungle, Hughes had worked in many countries across Asia and Africa as a freelance photographer taking pictures to support the work of mainly UK-based NGOs. As part of the practice he was always attuned to the power issues in play in this traditional approach to image gathering, however benign the intention. The ways he attempted to mitigate this was by photographing with consent, gathering names and quotes and asking, 'if I were the subject of this photograph, would I be happy to see it published?' When photographing politicians, military commanders and the like, these concerns were generally waived. Those wielding power can be legitimately held to account in visual or written media. Attempting to manage access to their portrayal is part of their work. The photographer must be wary of manipulation and try to avoid being co-opted into producing propaganda for those in power.

Pittaway, Bartolomei, and Hugman (2010) discuss the challenges of using participatory methods in their work with refugee research participants. They question the ethical responsibilities to the stories in research contexts while working with displaced populations, in complex and sometimes dangerous settings. The authors' main concern in posing the questions about what happens to the stories of participants lies in the ethical challenge for the researchers to contribute to the lives of people they work with, recognising them as subjects. Pittaway et al.

argue that power imbalances between researchers and participants raise complex questions while working with refugees, as the rights of research respondents who are protected by standard ethics procedure have been designed on the basis of biomedical research which usually take place in isolated, easy to control settings. Refugee participants do not generally have rights over their representation as research participants, or authority to make a complaint.

Anonymity is a particularly difficult issue in participatory photography. Richard Avedon spoke of the photographer and the subject having 'different ambitions for the image'. This transfers to participatory work, where an image may have different meanings and uses for the photographer, an academic researcher, an arts project, or a politically motivated advocacy group. Previously, when Hughes was working with school children on participatory photography, the project was framed as an art project and thus there was no suggestion that the children should be anonymised or instructed not to photograph themselves or other people. The idea of documenting a school with no children and families without human faces would have been perverse. Pupils would have photographed each other and their families regardless. The resulting policing of their photos would have been potentially emotionally damaging to the children. Explaining that the sight of their faces was inflammatory or obscene and that a photograph of them would place them in great jeopardy, would have been undermining and confusing. The fear and lack of confidence engendered could have reduced their ability to recognise and manage any issues around paedophilia they might encounter.

In other participatory projects Hughes has worked on, such as Through Positive Eyes,[2] with Gideon Mendel, involving those living with HIV, participants all agreed to photograph themselves recognisably. Where published sets of images did not feature faces this was for aesthetic, not ethical reasons. For example, Anthony in Mumbai used paper cut-out shadow figures to tell the story of his life with HIV. This was simply the strongest and most pared down method of re-visiting past trauma. Taking part in this project clearly exposed participants to the very real possibility of physical attack, abuse or discrimination. Each participant went through a thorough consent process. Volunteers were initially screened by local advocacy and support groups working with the projects. Most people had already openly disclosed their HIV status, and many were activists in the field. All agreed with the tenets of the project, that stigma was at least as dangerous to them as the virus itself and that if people saw their faces presented without shame and heard their stories in their own words then stigma would be reduced.

Risk of violence was also present with the participants in Calais, although participants were already clearly recognisable as refugees to all those who were exercising violence in the camp. The consequence of being recognised as a resident of the Jungle posed the biggest risk for those participants who had not (yet) sought asylum in France. Under the Dublin Agreement fingerprints can be held on record for five years before any records must be destroyed. Although some participants were planning to remain in France, they had to wait for a period of time after having been fingerprinted in Bulgaria, Greece, Serbia, or elsewhere before presenting their cases. Others were still hoping to move elsewhere in Europe, including the UK. For them the risk of being recognised from photos, especially any which were published, could have meant being returned to France after succeeding in the perilous journey across the English Channel. Some participants also wished to hide the fact they were living in the Jungle from their families, and indeed risk of violence towards families who stayed behind was something that could not be predicted, in cases of some participants, had they been shown to be in Europe. It also wasn't possible to predict whether all participants would be able to stay in Europe, and what the consequences might be if some had returned to their countries of origin with photographic evidence of their stay in Europe available publicly.

With due consideration and consultation with experts, we felt that the choice to be shown, where legal status had not yet been achieved, was too great to waive. The authorship of participants and their own choices of representation were severely limited by the fear of legal consequences for the participants. Participants remained the owners of their photos, and often posted them in social media accounts at the same time as the team was carefully removing them from other public sites. We were simultaneously trying to come up with ways to respect authors' choice of topics and representation while anonymising the pictures. Participants were guided to exclude faces from their photos and photograph backs, hands, shadows, and outlines instead. When a face was included, Hughes experimented with anonymising techniques. Both pixilating the face and covering it with a black bar seemed to draw parallels to representation of criminals in popular media and were immediately abandoned. Instead, Hughes faded the authors' faces from the image, making them appear transparent against the background. On reflection this seemed just as bad a solution. Authors, already rendered voiceless through media who presented them in a negative light, were once again erased from their own stories. While this technique did not evoke the same connotations of criminality as others, it rendered the refugee participants invisible, leaving them to haunt the beautiful European city in which they posed, which in real life was ignoring their lives and rights and erasing their humanity.

This contradictory and complex process of anonymising while respecting authorship and author choices of public display of photos could be seen as paranoid. Indeed, social media and photographs are rarely used as evidence in Dublin returns. To our best knowledge, no participants' details or images produced during the project were used against them in legal proceedings, and this renders the process worthwhile, even if it was indeed futile. Most of the authors have now gained safety and are able to reclaim their work from the project, if they so wish. With the help of the anonymity during the project, they can also leave the photographs, the stories, the project and the Jungle behind without the fear that it will come back to haunt them through these images. But only if the images themselves are also left behind, and do not continue to come back on the pages of newspapers, websites and methodology books.

CONSENT AND OWNERSHIP

Having discussed representation, power, and coloniality, we wished to prioritise ownership and consent throughout the project. Consent and ethics procedures in the social sciences have largely been lifted wholesale from the medical field.

In medical trials, there is a real need to anonymise data, not only to protect patient's medical records from being made public, but also to keep the trial blind and prevent bias on the part of the researchers.

We had avoided projects in which disposable film cameras were handed to NGO beneficiaries as being gestural and infantilising. All control would remain with the NGO and its foreign staff. The cameras implied that the beneficiaries' pictures would be low quality and disposable, and that they could not be trusted with a real camera. Disposable cameras are collected at the end of a project and kept by the organisers, who do not need to continue their relationship with the participant after this point. Participants may never get to see their own photos once they have been developed. This highlights the participants' own disposability, already at the forefront in the Jungle. Indeed, Hagan (2018) found that refugees in Calais were quick to turn down such an approach.

The advent of good quality, affordable, digital cameras presented an opportunity to properly up-end, or at least destabilise, the power relationship between photographer and subject. Ownership of the digital cameras and the memory cards, as well as the promise of a future ownership of these photos marked the participants as authors from the beginning. The issue of ownership is rather problematic in the context of the Jungle. In addition to the invisibility and lack of rights which frames the refugee experience in camps, the Jungle, as an unofficial camp, provided no security or right to property for the residents. Many residents left the camp every night to make the 7 km walk to the Calais port to 'try' for England. Anyone who didn't return within 12 hours was assumed to have left, both by the authorities and other residents. The container camp run by the French state imposed a 12-hour absence rule, after which individual's possessions would be removed and destroyed, monitored by automatic gates using a fingerprint as access key. Little could be 'owned' by anyone living in this situation. Hence, any discussion by us of participants 'owning' their photos was limited.

Crucial to our method is the editing process. Digital cameras offered a chance for subjects to become participants. If the images were not only taken by the participants but also seen, edited and captioned by them, then surely ethical and consent issues could be completely side-stepped. Levels of confidentiality, respect for privacy, notions of taste and style could all be set by the new photographers. The aesthetics of colonial inquiry, traditional anthropology, and ideas of the exotic which any photographer couldn't help bringing to bear on their subjects would all vanish in a puff of digital smoke. We cannot pretend that the images produced present the most plural and diverse range of truths that would be desirable. There were no female photographers and some participants had already been exposed to photography as an art form, and hence its traditional focus and preference for male gaze (Berger, 1972). Once it became apparent that some of these photos would be seen by wide audiences who could access them online as well as the readers of Voices from the Jungle, questions of the gaze, representation and authorship needed to be revisited. What kind of narrative were we (co-)producing if all we showed was squalor? Yet, as Dodd and Jackson (2019) argue, as editors of the book and the online exhibitions we were the ones to ultimately decide what 'good' photography from the Calais Jungle looked like.

When working in schools, Hughes and Mendel had involved the children in the editing at all stages. Each child had an initial one-to-one session to look at all their images. These images then went to a final group edit with the whole class. This developed their critical skills and self-esteem while promoting understanding of different lives and cultures within the school. The images, and short extracts written about them, including poems, were used and displayed in a variety of ways: exhibited in galleries, in multimedia videos, in curriculum development, and in print. In the Jungle, the opportunities for this level of participation were limited, but where possible, photos were selected with participants either in person, or at later stages via email.

Consent was discussed from the beginning of the project with participants, with each taking different levels of interest in the process. In addition to discussions regarding anonymity, participants were informed the photos would remain theirs, and consent would always be sought if photos were to be used. Consent and ownership continue to play out in complex ways outside of the camp. Pittaway et al. (2010, p. 233) identify a potential issue in the dissemination of the findings of research with refugee participants. That is, the desire of refugees themselves to have their stories told to the international community without considering potential harmful effects of those stories to themselves and their communities. Consent is also always given at a specific time. Here, that time was when participants were still living in the Jungle, or where they had recently left it. Once participants moved on and made lives elsewhere, the consent received at that time is no longer satisfactory. Assuming participants continue to wish to be represented as refugees and as residents of the once infamous refugee camp risks locking them into a static refugee identity.

We still believe that participatory photography is capable of producing far more challenging and interesting images than the traditional lone photographer. But the interplay of consent, power, distribution, control, and anonymity has become more complex. As a tool for academic research or for advocacy, it is now much less amenable to a tick-box ethics committee approach. The UK-based ethics committee is now attempting to police images which are no longer produced by its employees. If a refugee takes a heartfelt but sexist photograph, then the foreigners may step in and delete or suppress it – thus restoring the old power relation. In long-term educational projects, these questions can be addressed through dialogue, whereas in the temporary refugee camp setting the space for these dialogues was limited.

CONCLUSION: RELATIONALITY OF ETHICS

The camp was dismantled and the more than 8,000 people living there by October 2016 dispersed to locations around France. While the conditions in the camp had been inhumane, the demolition of it presented new problems. When dispersed into small communities, refugees have little access to cultural and educational initiatives. Those who felt they would not qualify for asylum in France have fled into other countries or are staying outside official accommodation. Calais has been a crossing point for the UK for many decades; people still arrive. Without the camp, there are approximately 500 refugees staying in Calais intermittently, trying to cross to the UK. They are harassed by the police at food collection points, kept from sleeping in large groups and are unable to establish any shelters or abodes, however temporary.

Listening became the most important ethical and decolonial practice in these educational and storytelling processes. If we understand life to provide materials for narratives, we can embrace the multiplicity of the possible narratives that can be told. As Meretoja (2018, p. 27) tells us, narrative is 'ethically loaded precisely because it is a way of making sense of our being in the world'. Our participants and co-authors continuously challenged our assumptions about ethics and representation. While we were at times overly concerned about the Dublin Accord and its consequences, our participants held insider knowledge about its inner workings that we did not have access to. For them, concerns about family back home were often more important; not wanting anyone to know they were 'living the Jungle life'.

Two years after the end of the project and publication of the book Voices from the Jungle, new challenges have emerged. On-going consent and difficulties to remain in touch with participants to reinstate this at every new public communication can stagnate discussions and dissemination. As participants have moved on with their lives, most with a significantly improved membership of their new European states, constantly asking them to return to the Jungle through these discussions is a new form of violence. Some are willing and eager to continue the conversations we started nearly four years ago. Others are not. Stigmatisation of migrants and refugees continues in public discussions and the media, and as our authors have been given an opportunity to move on, emailing or texting any of them, yet again, just asking if we can show that image again; include their name again; bring them back to the Jungle makes my heart heavy. For this reason, we have struggled with the decision of whether to include photos in this chapter. Contacting, out of the blue, asking for permission, discussing which photo to include, cannot be undone once we start the conversation. We become the ones to transport the original meaning to new, unknowable locations. We simultaneously transport all of us – ourselves, the authors and our audiences back to the Jungle, which now has gained a new meaning for all of us. Every time an invitation comes to go talk to a newspaper or a conference audience about 'what it was like', we wish to prioritise the voice of the participants. Every time, we also want to give the participants a chance to move on and leave the Jungle behind.

NOTE

1. Regular census was done in collaboration by L'Auberge des Migrants and Help Refugees in the camp. The official figures were reported as lower, and the reasons for this are discussed for example here https://fullfact.org/immigration/counting-number-migrants-calais-jungle/
2. Crispin Hughes and Gideon Mendel have worked with HIV+ve adults in a new worldwide project managed by the UCLA Art and Global Health Center. Despite its academic

base Through Positive Eyes was conceived from the start as an advocacy project combating the stigma of HIV: 'Through Positive Eyes is an attempt to address key themes of the AIDS epidemic: widespread stigma, extreme social inequality, and limited access to lifesaving medication. The project is based on the belief that challenging stigma against people living with HIV/AIDS is the most effective method for combating the epidemic – and that art is a powerful way to do this' (https://throughpositiveeyes.org/about).

REFERENCES

Andrews, M. (2014). *Narrative imagination and everyday life*. Oxford: Oxford University Press.

Benezer, G., & Zetter, R. (2014). Searching for directions: Conceptual and methodological challenges in researching refugee journeys. *Journal of Refugee Studies*, *28*(3), 297–318.

Berger, J. (1972). *Ways of seeing*. London: Penguin Books

Bignante, E. (2010). The use of photo elicitation in field research: Exploring Maasai representation and use of natural resources. *EchoGeo*, *10*. Retrieved from https://journals.openedition.org/echogeo/11622. Accessed on April 22, 2019.

Bruner, J. S. (2003). *Making stories: Law, literature and life*. New York, NY: Farrar, Straus and Giroux.

Butler, J. (2016). *Frames of war: When is life grievable?* London: Verso.

Calais Writers. (2017). *Voices from the "Jungle*." London: Pluto Press.

Collier, J. (1945). United States Indian administration as a laboratory of ethnic relations. *Social Research*, *12*(3), 265–303.

Corey, S. M. (1953). *Action research to improve school practices*. New York, NY: Bureau of Publications, Teachers College, Columbia University.

Dodd, S., & Jackson, A. (2019). "Good" photographs: The white male gaze and how we privilege ways of seeing. *Witness*. Retrieved from https://witness.worldpressphoto.org/good-photographs-the-white-male-gaze-and-how-we-privilege-ways-of-seeing-30ac3f005acc. Accessed on September 1, 2019.

Doloughan, F. (2006). Narratives of travel and the travelling concept of narrative: Genre blending and the art of transformation. *The Travelling Concept of Narrative*, *1*, 134–144.

Doná, G. (2007). The microphysics of participation in refugee research. *Journal of Refugee Studies*, *20*(2), 210–229. https://doi.org/10.1093/jrs/fem013

Drew, S. E., Duncan, R. E., & Sawyer, S. M. (2010). Visual storytelling: A beneficial but challenging method for health research with young people. *Qualitative Health Research*, *20*(12), 1677–1688. doi:10.1177/1049732310377455

Ellis, C. (2017). Compassionate research: Interviewing and storytelling from a relational ethics of care. In I. Goodson, M. Andrews, & A. Antikainen (Eds.), *The Routledge international handbook on narrative and life history* (pp. 431–445). New York, NY: Routledge.

Esin, C. (2017). Telling stories in the pictures: Constituting and processual and relational narratives in research with young British Muslim women in East London. *Forum Qualitative Sozialforschung/Forum: Qualitative Social Research*, *18*(1). Art. 15. Retrieved from http://www.qualitative-research.net/index.php/fqs/article/view/2774/4063

Esin, C., & Squire, C. (2013). Visual autobiographies in East London: Narratives of still images, interpersonal exchanges, and intrapersonal dialogues. *Forum Qualitative Sozialforschung/Forum: Qualitative Social Research*, *14*(2). Art. 1.

Etherington, K. (2007). Ethical research in reflexive relationships. *Qualitative Inquiry*, *13*(5), 599–616.

Finnish Institute in London and the Finnish Cultural Institute for the Benelux. (2016). Refugees and asylum seekers in press coverage. Retrieved from https://www.fininst.uk/wp-content/uploads/2017/11/Refugees_and_asylum_seekers_in_press_coverage.pdf. Accessed on April 22, 2019.

Freire, P. (1972). *Pedagogy of the oppressed*. New York, NY: Herder and Herder.

Gastaldo, D., Magalhães, L., Carrasco, C., & Davy, C. (2012). Body-map storytelling as research: Methodological considerations for telling the stories of undocumented workers through body mapping. Retrieved from http://www.migrationhealth.ca/undocumented-workers-ontario/body-mapping. Accessed on April 22, 2019.

Gauntlett, D., & Holzwarth, P. (2006). Creative and visual methods for exploring identities. *Visual Studies, Journal of the International Visual Sociology Association*, *21*(1), 82–91.

Gómez, R., & Vannini, S. (2015). *Fotohistorias: Participatory Photography and the Experience of Migration.* Seattle, WA: CreateSpace Independent Publishing.

Guillemin, M., & Gillam, L. (2004). Ethics, reflexivity, and "ethically important moments" in research. *Qualitative Inquiry*, *10*(2), 261–280.

Gomez, R., & Vannini, S. (2017). Notions of home and sense of belonging in the context of migration in a journey through participatory photography. *The Electronic Journal of Information Systems in Developing Countries*, *78*(1), 1–46.

Hagan, M. (2018). Capturing the invisible: The challenges of using photography as method with asylum seekers living in Calais, France. Retrieved from https://www.law.ox.ac.uk/research-subject-groups/centre-criminology/centreborder-criminologies/blog/2018/10/capturing. Accessed on April 22, 2019.

Houghton, A.-M., & Morrice, L. (2008). *Refugees, asylum seekers and migrants: Steps on the education and employment progression journey*. Leicester: National Institute of Adult Continuing Education.

Hugman, R., Pittaway, E., & Bartolomei, L. (2011). When "do no harm" is not enough: The ethics of research with refugees and other vulnerable groups. The *British Journal of Social Work*, *41*(7), 1271–1287.

Kofoed, J., & Staunæs, D. (2015). Hesitancy as ethics. *Reconceptualizing Educational Research Methodology*, *6*(2), 24–39.

Malone, B. (2015). Why Al Jazeera will not say Mediterranean "migrants". *Al Jazeera*. Retrieved from https://www.aljazeera.com/blogs/editors-blog/2015/08/al-jazeera-mediterranean-migrants-150820082226309.html. Accessed on April 5, 2019.

Meretoja, H. (2018). *The ethics of storytelling: Narrative hermeneutics, history, and the possible*. New York, NY: Oxford University Press.

Nayeri, D. (2019). *The ungrateful refugee*. London: Canongate.

Pittaway, E., Bartolomei, L., & Hugman, R. (2010) "Stop stealing our stories": The ethics of research with vulnerable groups. *Journal of Human Rights Practice*, *2*(2), 229–251.

Riessman, C. K. (2008). *Narrative methods for the human sciences*. London: Sage.

Rudduck, J. (1988). Changing the world of the classroom by understanding it: A review of some aspects of the work of Lawrence Stenhouse. *Journal of Curriculum and Supervision*, *4*(1), 30–42.

Sanyal, D. (2017). Calais's "Jungle" Refugees, Biopolitics, and the Arts of Resistance, Representations, 139(1), 1–33. doi: 10.1525/rep.2017.139.1.1.

CHAPTER 2

USING PARTICIPANT-PRODUCED DRAWINGS AS AN ALTERNATIVE TO PHOTOGRAPHS IN ETHNOGRAPHIC RESEARCH

Angela Stephanie Mazzetti

ABSTRACT

In this chapter, the author explores the ethical challenge of preserving participant anonymity when using visual methods in ethnographic research. Referring to her own ethnographic study in post-conflict Northern Ireland, the author explores how social, cultural, and political contexts may accentuate the need to preserve anonymity. The author discusses her rationale for opting not to use photographs in this context and puts forward the case for using participant-produced drawings as an alternative to photographs. Drawings accomplish similar rich benefits as photographs but may ameliorate the ethical challenges inherent in photographic work of maintaining participant anonymity.

Keywords: Visual methods; Northern Ireland; sensitive research; participant-produced drawing; drawing; ethnography

INTRODUCTION

The image below (Fig. 1) is the result of my lack of technical knowhow as my flash reflects off the glass picture frame I am attempting to photograph. Although my face is obscured by the camera, my identity is unwittingly revealed as my

Ethics and Integrity in Visual Research Methods:
Advances in Research Ethics and Integrity, Volume 5, 29–42

ISSN: 2398-6018/doi:10.1108/S2398-601820200000005005

Fig. 1. Linen Hall Library Belfast, February 2017. Photographed by Angela Stephanie Mazzetti.

very distinctive coat and hair colour may potentially reveal my identity to those who know me, have met me, or who have seen other published images of me. And yet, preserving anonymity is a key principle underpinning ethical research in the social sciences (Wiles, Crow, Heath, & Charles, 2007). Addressing this issue requires that researchers do not disclose participants' identities and also that researchers fully brief their research participants as to the extent to which anonymity can or cannot be assured in the dissemination and publication of research outputs (Wiles et al., 2008).

The use of photographs, specifically those produced by research participants, presents researchers with a number of ethical challenges regarding both participant anonymity and data integrity (see Wiles et al., 2008 for a more comprehensive review of the issues). First, photographs taken by participants may present researchers with images that capture people in public spaces who have not given their consent for their image to be used (Wiles et al., 2008) or participants may capture images of people engaged in unlawful or immoral activities (Clark, Prosser, & Wiles, 2010). Furthermore, although participants may take photographs for the specific purposes of the research study to which they have been recruited, they may personally use the images for their own purposes which may involve the unrestricted sharing of images on social media platforms without peoples' identities being concealed (Wiles et al., 2008).

Second, photographs taken by research participants may present researchers with issues of internal confidentiality (Pink, 2007; Wiles et al., 2008). In the context of ethnographic research, researchers are working within defined, often small communities. As such, photographs taken within these communities (whether taken by the researcher or the participants), may render members of those communities easily recognisable to one another (Banks, 2007; Rose, 2007; Surmiak, 2018). Even when photographs have been manipulated to disguise the faces of individuals (e.g. through blurring or pixilation), participants may still be recognisable to others through their clothes, distinctive jewellery, or physical features such as tattoos (Wiles et al., 2008). Furthermore, contextual background information (such as a street name or a building) may also unwittingly reveal identities (Prosser & Loxley, 2008). As such, using photographs as a method of data capture may render the process of anonymity (of people and places) more challenging as there is a greater likelihood of identification (Clark et al., 2010; Wiles et al., 2008).

In this chapter, I discuss these ethical challenges by reflecting on my own research experiences of using visual methods. In particular, I refer to my recent ethnographic study of stress appraisal and coping in post-conflict Northern Ireland, a study in which I decided to use participant-produced drawings as an alternative to participant-produced photographs. This decision was strongly influenced by the social, cultural, and political context of my research; a context in which I considered that the taking of photographs had the potential to inflict harm (in both the short-term and the long-term) on my research participants. This chapter is structured in four parts. First, I present an overview of the general ethical challenges of using visual methods in ethnographic research. Second, I provide an overview of my research context and how this context accentuated my desire to protect my research participants' identities. Third, I put forward the rationale for using participant-produced drawings as an alternative to photographs. Finally, based on my own research experiences, I put forward some suggestions that other researchers engaged in similar research may find useful.

THE ETHICAL CHALLENGES OF USING PARTICIPANT-PRODUCED PHOTOGRAPHS IN ETHNOGRAPHIC RESEARCH

Ethnography is the study of people in their normal social and cultural settings with the objective of establishing the meanings that people assign to their everyday activities (Hammersley & Atkinson, 2007). The role of the ethnographer is to bring their participants' worlds 'to life' for those unfamiliar with the research setting (Madden, 2017); a process that relies on a communicative trust and dialogue between ethnographer and participants (O'Reilly, 2012). Visual methods are a long-established method of data collection in ethnographic research (Emmison & Smith, 2000; Pink, 2007) complementing other ethnographic research methods such as observations and interviews (Prosser & Loxley, 2008). Visual methods encompass the use of photographs, video, internet pages, drawings, and visual representations (Prosser & Loxley, 2008). Photographs as a data collection method have been widely used in ethnographic research as visual imagery is seen as being central to our perceptions and interpretations of our everyday social lives (Pauwels, 2008). As such, photographs are a powerful means of capturing meaning and sense-making as they often reveal more about our participants' worlds than simply textual data alone (Collier & Collier, 1986; Pink, 2007). Photographs comprise 'found data' (e.g. archival photographs or family albums), researcher-produced data (photographs taken by the researcher), and participant-produced data (photographs taken by research participants for the specific purposes of the research project) (Prosser & Loxley, 2008, p. 5).

Contemporary approaches to ethnography advocate collaborative research between ethnographer and participants in the co-production of data, a move away from traditional, colonial, and paternalistic approaches to ethnography which placed the researcher as the independent and objective observer of the life of others (Pink, 2007). From this traditional perspective, the visual was used as a means of supporting ethnographers' claims and interpretations of the lives of others (Collier & Collier, 1986). A contemporary and collaborative approach to ethnography encourages a mutually trusting relationship between participants and researcher (Banks, 2001; Pink, 2007), which facilitates participant agency and voice, empowering participants to present their world as they see it and experience it (Allen, 2015; Banks, 2001; Pink, 2007). As such, participant-produced photographs have become widely used in ethnographic research as a means for participants to share their perspectives of their worlds (Allen, 2015). However, this paradigm shift in the use of photography raises a number of ethical challenges for ethnographic researchers specifically related to participant anonymity and data integrity.

Obscuring photographs through digital manipulation (such as pixilating, cropping, or applying black-out bars to faces), before dissemination and publication, is a common way that researchers anonymise photographs to protect the identity of research participants (Allen, 2015; Wiles, Coffey, Robinson, & Heath, 2012a). However, altering images is problematic. First, tampering with an image may compromise the integrity of the data (Allen, 2015; Pauwels, 2008; Wiles et al., 2012a). Afterall, ethnographers adopt visual methods because images have the potential to portray richer contextual information than textual data alone (Banks, 2001; Pink, 2007; Rose, 2007). As such, tampering with an image may erase this essential contextual information that it set out to capture in the first place (Prosser & Loxley, 2008; Wiles et al., 2012a). For example, obscuring a person's face also obscures the meaning behind the image, resulting in the loss of an essential facet of the image (Allen, 2015). Afterall, what value does a blurred face add to an interpretation of meaning? (Clark et al., 2010). As such, obscuring an image results in important nonverbal information such as facial expressions (Pauwels, 2008) or emotion and feeling (Wiles et al., 2008) being lost.

Second, contextual or background information may render obscuring the face pointless (Clark et al., 2010; Pauwels, 2008). For example, reflecting on his 'connected lives' project, Andrew Clark (cited in Wiles et al., 2008) highlights that even though he had anonymised both his participants and the name of the field site, those who were familiar with the research setting were able to recognise themselves and others in the photographs. Third, obscuring facial features is a contentious issue in the social sciences, (Sweetman, 2009) juxtaposing on the one hand, participant empowerment and voice, and on the other, researcher paternalism (Allen, 2015). Individuals and communities may want to be identified and therefore obscuring their faces is disconcerting, disrespectful, and objectifying (Allen, 2015; Wiles et al., 2008). Furthermore, the masking of the face may have negative social connotations associated with crime or victimhood (Wiles et al., 2008).

However, Pink (2007) argues that there is a tension between participants' rights to decide how their images will be used and the ethnographer's responsibility to inform participants of the implications of their decisions. She highlights that participants may not be cognisant of the norms and conventions of the research world and therefore they need to be informed of the potential outlets (such as conference papers, books, and visual displays) where their images might be shared. In an era of open access, there is an additional risk that photographs will be further distributed, distorted, and misrepresented outside of their original research setting (Mannay, 2014) as they are shared and re-used through cyberspace (Prosser & Loxley, 2008). Once an image enters the public domain, it is difficult to remove it (Banks, 2001). As such, Pink (2007) stresses that researchers need to take responsibility for the outcomes of their research and in certain contexts, this may involve anonymising images to protect themselves and their participants.

Finally, ethnographers need to be aware of the social, cultural, and political contexts in which their photographs may be viewed and interpreted (Pink, 2007; Prosser & Loxley, 2008). Such contexts will inform research ethical decision-making about how, when and if photographs should be taken and published (Langmann & Pick, 2014; Pink, 2007; Wiles, Coffey, Robinson, & Prosser, 2012b). As such, researchers need to be sensitive to the situational demands of the research setting and be responsive and adaptable in addressing ongoing ethical challenges (Browne & Moffett, 2014). Guillemin and Gillam (2004) make the distinction between procedural ethics (the practices mandated by ethical review panels) and ethics in practice (the ongoing ethical concerns that researchers face when in the field). Lester and Anders (2018) suggest that procedural ethics may fail to anticipate the ethical dilemmas encountered in visual research and as such, researchers need to make ongoing decisions about what is ethically correct in a given context. This has specific significance in the context of sensitive research[1] (Lee, 1993), research involving vulnerable groups (Surmiak, 2018), or research in conflict and post-conflict societies (Browne & Moffett, 2014; Dawson, 2007; Ganiel, 2013).

THE IMPORTANCE OF MAINTAINING PARTICIPANT ANONYMITY IN ETHNOGRAPHIC RESEARCH

Between July 2016 and December 2018, I conducted an ethnographic study with a group of participants who had grown up in what can be categorised as the

Belfast Metropolitan Area.[2] The purpose of the study was to explore the long-term emotional impact of coping with the Troubles. Many lives in Northern Ireland were affected by the Troubles, the term euphemistically given to a 30-year period of intra-state conflict which lasted from the civil unrest and mobilisation of British troops in 1969 until the signing of the Good Friday/Belfast Agreement in 1998 (Dixon & O'Kane, 2011; Dunn, 1995). During the Troubles, over 3,700 people were killed and over 40,000 were injured (Smyth & Hamilton, 2003). In a small geographical area with a relatively small population, it is estimated that one in two people lost a friend or acquaintance to the violence and one in ten, a family member (Gallagher, 2014). The protracted and vicious nature of the conflict has left deep emotional scars (Bolton, 2017; Dawson, 2007; McKay, 2008) and even after twenty years since the signing of the Good Friday/Belfast Agreement, the emotional legacy of the Troubles continues to impact the region as Northern Ireland attempts to come to terms with its past and build a peaceful future (Dawson, 2007).

Although I considered that my research was not political with a capital P (Ganiel, 2013), the situational and temporal context of my research cast a political shadow over my research. My fieldwork took place during a volatile political period in Northern Ireland which saw the marginalisation of the middle of the road political parties, the collapse of the Northern Ireland Assembly, Brexit and with it, the threat of the reintroduction of a 'hard border' between Northern Ireland and the Republic of Ireland. In addition, my fieldwork coincided with commemorative events of significance to the political landscape of Northern Ireland, such as the fiftieth anniversary of the civil rights marches, the twentieth anniversary of the signing of the Belfast/Good Friday Agreement, and the centenary anniversaries of both the Battle of the Somme and the Easter Risings.

Dixon and O'Kane (2011) highlight that, although there is consensus that the region is now more peaceful, what remains contested is how secure that peace is. As one of my research participants noted, 'old wounds never really heal in Northern Ireland, they just fester under the surface, and each time they break open, they are harder to repair'. In this temporal and situational context, talking about the Troubles continues to be a controversial and contentious issue. There are different narratives as to the cause of the conflict, different narratives as to who should be considered the victims of the conflict, and different narratives as to how to maintain peace in a society with a perceived hardening of sectarian attitudes, ongoing challenges of curbing dissident paramilitary activity, and dealing with the sectarian sticking points (such as flags and parades) (Dawson, 2007; Dixon & O'Kane, 2011). As such, my study was sensitive both in terms of the intrusive nature of the research topic (emotion, stress, and coping) but also in terms of the political sensitivities of conducting research in a post-conflict society.

A photograph captures someone's identity in a way that other data collection methods cannot access (Langmann & Pick, 2014) and throughout my research in Northern Ireland, I was frequently reminded of the culture of fear, mistrust, secrecy, and surveillance that still permeates much of Northern Irish society. As an illustration, on my first fieldwork visit to Northern Ireland, I met with an organisation that had used participant-produced photographs to complement participant interviews in their book collection of Troubles-related narratives. However, the original edition of the book had to be destroyed as one of the families had retrospectively decided not to have their story or image made public. This resulted in a revised edition not being available for the book launch and therefore pages had to be physically ripped from the original edition so as to omit the family's story. Later in my fieldwork, I encountered a similar situation with another organisation who had likewise used participant-produced photographs to complement participant interviews. On this occasion, two participants asked for their narratives and pictures to be removed in the final stages of getting the collection ready for publication. One participant was so concerned about the repercussions that the participant asked that all emails, communications, drafts of the story, and photographs be permanently deleted from the organisation's communications systems. Furthermore, at the end of my fieldwork, I encountered a social media news alert relating to an incident in Belfast in which a man had been attacked. The article explicitly referenced the fact that the victim of the attack wanted his identity preserved. The original article included a picture of the victim to illustrate the extent of his injuries, but to protect his identity, the face of the victim was pixelated. However, on subsequent shares of the article on social media, the original pixilation had been removed and the man's face was revealed.[3]

The examples above highlight how the situational context may influence ongoing willingness of research participants to engage with a study. Ethnographic research is a long-term and unpredictable process (Pauwels, 2008). In the context of my research, the changing political climate in the region evoked participants' fears about being identified. Wiles et al. (2008) note that although individuals may be happy for their images to be taken at one point in their lives, they may be less happy if their situational circumstances change. The safety and dignity of research participants is paramount and as such, researchers must be attuned to circumstances under which real harm may be inflicted to participants through the use of visual data (Langmann & Pick, 2014; Pauwels, 2008). Clark et al. (2010) therefore suggest that researchers need to be aware of their situational contexts and apply this awareness to decide when and when not to use photographs.

This contextual decision-making needs to be embedded throughout the research process, from design through to dissemination and publication (Pauwels, 2008). With regards to my own study, this situational awareness was embedded throughout. For example, even before I commenced the fieldwork phase of my research, I attended an advanced training programme on the ethics of conflict and peace research. This course offered me the opportunity to engage in meaningful debate and discussion on issues related to ethical research in transitional contexts, including the use of photographs. Additionally, I also engaged in pre-fieldwork exposure visits to familiarise myself with the research context. Browne and Moffett (2014) suggest that pre-fieldwork exposure helps to develop researcher experience, situational awareness and confidence. As part of this pre-fieldwork exposure, as well as visiting Northern Ireland on a regular basis, I also kept up to date with the day-to-day climate in the region by subscribing to regional news alerts. This situational exposure was instrumental in my decision not to use participant-produced photographs or researcher-produced photographs of my participants or their immediate environments from the start of my fieldwork. I considered that revealing my participants' identities in this manner had the potential to inflict harm both in the short-term and in the longer-term. Finally, my decision not to use photographs was subsequently reinforced through my fieldwork encounters, as outlined above. Pauwels (2008) warns against a hit and run mentality to the use of visual methods, stressing that if researchers are aware of potential harm to their participants, then regardless of how valuable the method may be, it should not be used.

ADVOCATING THE USE OF PARTICIPANT-PRODUCED DRAWING IN POST-CONFLICT ETHNOGRAPHIC RESEARCH

Drawing is an ideal visual method for producing data on complex, emotive, and multi-faceted issues that are difficult to articulate verbally (Bagnoli, 2009; Prosser & Loxley, 2008) such as the topic of stress and coping (Mazzetti & Blenkinsopp, 2012). This is because drawing provides an outlet for deeply embedded memories and thoughts which may not be surfaced through other data collection methods such as interviews (Vince, 1995; Vince & Brousinne, 1996). Drawing, as a technique, may also liberate participants from cultural taboos about what can be said (Vince, 1995) and as such, drawing may elicit more honest expressions of emotions, feelings, assumptions, beliefs, and biases (Nossiter & Biberman, 1990). Drawing can facilitate the sense-making of complex transitions (such as post-conflict transitions) as it enables participants to synthesise events (Barner, 2008) and bring together interrelated memories and experiences (Mazzetti & Blenkinsopp, 2012). Furthermore, engaging participants in drawing fulfils the ethnographic principles of collaborative research (Pink, 2007) and empowers participants to both determine the content and the voice of the image they produce independent of the researcher (Banks, 2001). As such, participant-produced drawing as a data collection method, elicits similar benefits to participant-produced photographs. However, drawings (as opposed to photographs) ameliorate the ethical challenge of exposing participants' identities. To illustrate this point, I refer to a participant-produced drawing from my own research. The drawing was produced by the participant after a conversation we had on childhood memories and how these memories inform present-day perceptions of stress.

> I would dream about a fictional 'assault' on our house. Feeling like outsiders, it was easy to imagine that there was an unseen resentment against our family. A situation would occur in my mind where a gang of 'bad people' would attack our house. We had a room in the attic and the story would revolve around our defence of this space. There was a skylight that served as an escape route onto the roof where I would somehow manage to get onto street level. As the situation became dire and defeat imminent, a group of 'friends' would appear on the crest of the bridge and they would charge to our rescue. This would play out like an action movie in my mind. I would rework it over time, adding or adjusting different details and imagining different plot twists and outcomes. However, the story never seemed to reach a natural conclusion as I would awaken before it ended. (Personal communication)

The drawing of the 'Skylight' presented below (Fig. 2) (along with the participant's interpretation of its meaning and significance) enabled my participant to present a difficult memory from his/her childhood; that of feeling marginalised and under threat of attack, but without revealing his/her identity. For example, biographical data about the participant (in terms of gender, age, or religious affiliation) are not explicit in either the drawing or the participant's explanation of the drawing's meaning. Additionally, although the drawing is of a 'place' and the explanation refers to 'place', there are no location identifiers obvious in either the drawing or the explanation. This is particularly important in ethnographic research within small communities as a photograph of 'place' can easily identify the community to which the participant belongs (Wiles et al., 2008). Finally, the drawing makes no reference to insignia, emblems or colours (such as flags or murals), which are of particular significance in Northern Ireland (Jarman, 1999). As such, the drawing is not used to perpetuate a particular stereotype or narrative regarding the place (Wiles et al., 2008).

Fig. 2. 'The Skylight', an Original Participant-produced Drawing.

However, drawings are not without their own specific practical and ethical challenges. For example, some participants may feel uncomfortable or self-conscious about drawing (Mazzetti & Blenkinsopp, 2012). It is important therefore that the researcher makes it clear that the use of drawing is not a test of participants' artistic skills, but rather a way of empowering participants to visualise rather than verbalise sense-making and meaning (Bagnoli, 2009). The drawing presented above is rather sophisticated in technique and was produced outside of the research interview. As such, the participant was able to take his/her time in putting together the image without any interference from me as the researcher. However, other participants produced less sophisticated but no less meaningful and powerful drawings.

In this study, the participants were able to draw their own images in their own space and own time using whatever artistic medium they preferred. Drawings can also be used within the context of a research interview, however, this approach may put time pressures on the creative process and also restrict the medium to those made available by the researcher. This was an issue I encountered in an earlier research study in which I used visual timelines to explore career transitions. Many of the participants expressed reservations regarding their artistic prowess. I dealt with this issue by encouraging those less confident in drawing, to use emoticons to express their feelings. These emoticons were often embellished to emphasise emotions. For example, adding tears to a sad emoticon or adding a sunshine halo to a happy emoticon. Hybrid emoticons were used by some to express conflicting or mixed emotions, and varying sizes of emoticons were used to illustrate different levels of emotional responses or to symbolise growing or diminishing emotions over time (see Mazzetti & Blenkinsopp, 2012, for a detailed discussion). As such, even those who were unsure about their ability to be artistic were still able to draw insightful images.

Furthermore, ethical issues may arise from drawings if participants choose to include personal or place identifiers within the images (Wiles et al., 2008). It is therefore advisable that researchers encourage participants not to include such identifiers (e.g. peoples' names or place names, initials, dates) that may later need to be obscured when the drawings are shared or published. Referring to her Reflection on Young Lives project (cited in Wiles et al., 2008), Bagnoli discusses how she needed to alter the participant-produced images (self-portraits, collages, timelines, and relational maps) prior to dissemination of the images to remove any identifiers. She also reflects on the aesthetic challenges this presented in terms of giving voice to the young people involved in her study and their artistic efforts whilst at the same time tampering with and changing their artworks to preserve their anonymity. This was also a challenge I encountered in my timelines study. Many of the participants of that study included identifiers in their drawings, presenting me with ethical, technical, and financial challenges when disseminating and publishing the drawings. For example, the image used in Mazzetti and Blenkinsopp (2012) required the input of a digital technician with specialist software to remove identifiers (such as names and dates) from the timeline so that the image could be published without revealing the participant's identity. Furthermore, the size of the drawings in the timelines study (A1 sheets of paper) created difficulties with the storage and subsequently, the disposal of the images which required the use of a large format shredder, adding additional costs to the project (see Mazzetti, 2014 for a detailed discussion of the issues).

CONCLUSIONS

In this chapter, I provide an insight into my decision not to use photographs of my participants or participant-produced photographs. These decisions were driven by the social, political, and cultural context of my research. I considered that using photographs in a post-conflict society had too much potential to harm my participants through the disclosure of their identity. However, I also explore the value of visual methods as they have the capacity to bring rich and multi-faceted insights that cannot be achieved through textual data alone. I, therefore, explore my rationale for the use of participant-produced drawings as an alternative to photographs. I consider that participant-produced drawings elicit many of the benefits of photographs but ameliorate many of the ethical challenges. I also consider that these insights may be transferable to other research contexts. Prosser and Loxley (2008) note that there is a paucity of case study examples of how researchers adopting visual methods make ethically informed decisions. As such, I conclude with a set of ethical considerations that researchers in similar contexts may wish to consider as part of their ethical decision-making both in context and in practice.

Gaining ethical clearance: At the research design phase, it is important that researchers consider the rationale for adopting visual methods. It is worth contemplating at this stage the value that visual data will bring to the research project and how visual data will be captured (e.g. through photo-elicitation, researcher-produced, or participant-produced images). If the decision is taken to adopt researcher or participant-produced photographs, then consideration must be given to issues of informed consent and anonymity. This decision will require an understanding of the social, cultural, and political contexts in which the images will be created, interpreted, and disseminated (Pink, 2007). Pre-exposure to the research setting or prior knowledge and understanding of the research setting will support a more informed decision-making process on these issues.

In addition, ethics review panels will need to be presented with a sound rationale for the approach taken along with information on how issues of informed consent and anonymity will be dealt with. This will generally require that all stages of the research process have been considered including data capture, data storage, data sharing, and data dissemination and publication. Even after ethical clearance has been approved, researchers need to be attuned to the tensions between procedural ethics and ethics in practice (Guillemin & Gillam, 2004), and, as such, researchers will have to make ongoing ethical decisions in the field. The support of a research buddy, supervisor, or director with whom these challenges can be discussed, can facilitate this process. Issues of consent and anonymity are not just appropriate to photographs but also drawings. Additionally, researchers will need to have considered how they will manage the ongoing issues of participant anonymity throughout all stages of the research process.

Informed participants: It is important that researchers reflect on how they will inform research participants as to how their photographs or drawings will be used. This should include information on how these artefacts will be disseminated and published. It is useful to demonstrate this in practice, for example, by showing participants some examples of how their images will be used. This can involve sharing with them a poster, presentation, or publication that has used visuals. This process facilitates a better understanding of the journey that their images will take. It is also worthwhile discussing the timeframes associated with this. For example, it may take a few years for their images to be published and participants need to be informed that they are consenting to a long-term process and, at times, quite a dynamic process. Additionally, participants need to be informed that, even after the original artefact has been destroyed, digital images will remain in the public domain *infinitum* (Banks, 2001). In an age of open access, their images may also be used for further research purposes (Mannay, 2014).

Informed method: The process of data capture must be fully explained to participants. For studies that involve participant-produced photographs, this explanation needs to make reference to what is permissible to photograph and what is off limits. Furthermore, consideration needs to be given to the social, cultural and political context in which the research is situated as different contexts may inform the evaluation of what is acceptable and what is not acceptable to photograph (Pink, 2007). For studies that involve participant-produced drawings, researchers should consider how they will deal with participants' anxieties regarding their ability to draw. This may involve researchers providing different approaches to drawing such as collages, interpretive art, use of colour, self-portraits, timelines, or emoticons (Bagnoli, 2009; Mazzetti & Blenkinsopp, 2012; Wiles et al., 2008). Researchers may also wish to consider the group production of art that enables participants to work together in teams to co-produce drawings (Barner, 2008).

Anonymising images: Researchers need to reflect on how images will be anonymised. If identifiers are included (regardless of whether photographs or drawings are used) then retrospectively removing identifiers can require technical knowledge and specialist software and may also add additional costs to the research. Thought also needs to be given to how permanent any image manipulation technique may be. For example, the ease with which methods of digital anonymisation (such as blurring or pixilation) can be removed and then re-shared without anonymisation by those with the requisite technical knowledge. Thought also needs to be given to how the images will be used in the future and this may involve a decision not to include some aspects of visual data (Allen, 2015). Furthermore, in research studies that involve defined groups, internal confidentiality needs to be considered as individuals and groups may be easily recognisable to one another in such contexts (Prosser & Loxley, 2008).

Participant-informed analysis: Data integrity is an important principle underpinning social sciences research (Clark et al., 2010). The value of using participant-produced visual images is based on the premise that the method empowers participants to present their world as they see it and experience it (Pink, 2007). In deciding to adopt participant-produced images, it is therefore important that researchers consider how they will capture their participants' interpretation of these images. This can be achieved in a number of ways. First, the image may be produced as part of a research interview as in the timelines example discussed above. As part of this process, annotations (made by both the researcher and the participant) were added to the timelines to capture participant meaning throughout the process and also at the end of the interview (see Mazzetti, 2014, p. 9 for a detailed discussion of the process). Second, the participant-produced images may be discussed as part of a research interview, which is recorded and transcribed, to elicit participant meaning (Allen, 2015). Or, participants can add a textual explanation (outside of a research interview setting) to accompany their image. This method was adopted in this study and both the participant-produced drawing and the participant-produced interpretation of its meaning are presented above.

NOTES

1. Lee (1993, p. 3) defines sensitive research as research that has the potential to be intrusive (e.g. the study of emotion-laden topics); incriminating (such as the study of deviant behaviours); or political (e.g. studies of coercion or power struggles). Lee further highlights that 'sensitivity' is situationally and culturally defined and as such, 'normal' research may trespass into sensitive territory as the situational context of the research evolves.
2. An area that extends from Belfast city centre southwards to Lisburn and northwards to Carrickfergus on the Antrim coast, and Bangor on the Down coast.
3. I have decided not to include a picture from the news article as to do so, would perpetuate the matter.

REFERENCES

Allen, L. (2015). Losing face? Photo-anonymisation and visual research integrity. *Visual Studies, 30*(3), 295–308. doi:10.1080/1472586X.2015.1016741

Bagnoli, A. (2009). Beyond the standard interview: The use of graphic elicitation and arts-based methods. *Qualitative Research, 9*, 547–570. doi:10.1177/1468794109343625

Banks, M. (2001). *Visual methods in social research*. London: Sage.

Banks, M. (2007). *Using visual data in qualitative research*. London: Sage.

Barner, R. (2008). The dark tower: Using visual metaphors to facilitate emotional expression during organizational change. *Journal of Organizational Change Management, 21*, 120–137. doi:10.1108/09534810810847075

Bolton, D. (2017). *Conflict, peace and mental health: Addressing the consequences of conflict and trauma in Northern Ireland*. Manchester: Manchester University Press.

Browne, B., & Moffett, L. (2014). Finding your feet in the field: Critical reflections on early career researchers on field research in transitional societies. *Journal of Human Rights Practice, 6*(2), 223–237. doi:10.1093/jhuman/huu010

Clark, A, Prosser, J., & Wiles, R. (2010). Ethical issues in image-based research. *Arts and Health, 2*(1), 81–93. doi:10.1080/17533010903495298

Collier, J., & Collier, M. (1986). *Visual anthropology: Photography as a research method*. Albuquerque, NM: University of New Mexico.

Dawson, G. (2007). *Making peace with the past? Memory, trauma and the Irish troubles*. Manchester: Manchester University Press.

Dixon, P., & O'Kane, E. (2011). *Northern Ireland since 1969*. Harlow: Pearson Education Limited.

Dunn, S. (1995). *Facets of the conflict in Northern Ireland*. Hampshire: McMillan Press Ltd.

Emmison, M., & Smith, P. (2000). *Researching the visual*. London: Sage.

Gallagher, A. M. (2014). Psychological approaches to the Northern Ireland conflict. *Canadian Journal of Irish Studies, 13*(2), 21–32. doi:10.2307/25512706

Ganiel, G. (2013). Research ehics in divided and violent societies: Seizing the ethical opportunity. In C. Russell, L. Hogan & M. Junker-Kenny (Eds.), *Ethics for Graduate Researchers* (pp. 167–181). Elsevier. doi:10.1016/B978-0-12-416049-1.00011-8

Guillemin, M., & Gillam, L. (2004). Ethics, reflexivity and ethically important moments in research. *Qualitative Inquiry*, *10*(2), 261–280. doi:10.1177/1077800403262360

Hammersley, M., & Atkinson, P. (2007). *Ethnography*. New York, NY: Routledge.

Jarman, N. (1999). Commemorating 1916, Celebrating difference: Parading and painting in Belfast. In A. Forty & S. Küchler (Eds.), *The art of forgetting* (pp. 171–195). Oxford: Berg.

Langmann, S., & Pick, D. (2014). Dignity and ethics in research photography. *International Journal of Social Research Methodology*, *17*(6), 709–721. doi:10.1080/136455792013.825473

Lee, R. (1993). *Doing research on sensitive topics*. London: Sage.

Lester, J. N., & Anders, A. D. (2018). Engaging ethics in postcritical ethnography: Troubling transparency, trustworthiness, and advocacy. *Forum: Qualitative Social Research*, *19*(3), 1–18.

Madden, R. (2017). *Being Ethnographic: A Guide to the Theory and Practice of Ethnography.* London: Sage Publications Ltd.

Mannay, D. (2014). Storytelling beyond the academy: Exploring roles, responsibilities and regulations in the open access dissemination of research outputs and visual data. *Journal of Corporate Citizenship*, *54*, 109–116.

McKay, S. (2008). *Bear in mind these dead*. London: Faber and Faber Ltd.

Mazzetti, A. (2018). *To Tape or Not to Tape: Reflections on the Complexities of Taping Ethnographic Interviews.* SAGE Research Methods Series. London: Sage Publications Ltd.

Mazzetti, A., & Blenkinsopp, J. (2012). Evaluating a visual timeline methodology for appraisal and coping research. *Journal of Occupational and Organizational Psychology*, *85*(4), 649–665.

Nossiter, V., & Biberman, G. (1990). Projective drawings and metaphor: Analysis of organizational culture. *Journal of Managerial Psychology*, *5*, 13–16.

O'Reilly, K. (2012). *Key concepts in ethnography*. London: Sage.

Pauwels L. (2008). Taking and using. *Visual Communications Quarterly*, *15*(4), 243–257. doi:10.1080/15551390802415071

Pink, S. (2007). Doing *visual ethnography*. London: Sage.

Prosser, J., & Loxley, A. (2008). *Introducing visual methods*. ESRC National Centre for Research Methods Review Paper. Retrieved from http://eprints.ncrm.ac.uk/420/1/MethodsReviewPaperNCRM-010.pdf.

Rose, G. (2007). *Visual methodologies*. London: Sage

Smyth, M., & Hamilton, J. (2003). The human costs of the troubles. In O. Hargie & D. Dickson (Eds.), *Researching the troubles: Social science perspectives on the Northern Ireland conflict* (pp. 15–36). Edinburgh: Mainstream Publishing.

Surmiak, A. (2018). Confidentiality in qualitative research involving vulnerable participants: Researchers' perspectives. *Forum: Qualitative Social Research*, *19*(3), 1–11.

Sweetman, P. (2009). Revealing habitus, illuminating practice: Bourdieu, photography and visual methods. *The Sociological Review*, *57*, 491–511. doi:10.1111/j.1467-954X.2009.01851.X

Vince, R. (1995). Working with emotions in the change process: Using drawings for team diagnosis and development. *Organisations and People*, *2*, 11–17.

Vince, R., & Broussine, M. (1996). Paradox, defense and attachment: Accessing and working with emotions and relations underlying organisational change. *Organization Studies*, *17*, 1–21.

Wiles, R., Coffey, A., Robinson, J., & Heath, S. (2012a). Anonymisation and visual images: Issues of respect, 'voice' and protection. *International Journal of Social Research Methodology*, *15*(1), 41–53. doi:10.1080/13645579.2011.564423

Wiles, R., Coffey, A., Robinson, J., & Prosser, J. (2012b). Ethical regulation and visual methods: Making visual research impossible or developing good practice? *Sociological Research Online*, *17*(1), 1–10. doi:10.5153/sro2274

Wiles, R., Crow, G., Heath, S., & Charles, V. (2007). The management of confidentiality and anonymity in social research. *International Journal of Social Research Methodology*, *11*(5), 417–428. doi:10.1080/13645570701622231

Wiles, R., Prosser, J., Bagnoli, A., Clark, A., Davies, K., Holland, S., & Renold, E. (2008). Visual ethics: Ethical issues in visual research. *ESRC National Centre for Research Methods*, *11*. Retrieved from http://eprints.ncrm.ac.uk/421/1/MethodsReviewPaperNCRM-011.pdf

CHAPTER 3

THE PEOPLE IN THE PICTURES RESEARCH: TAKING CARE WITH PHOTO ELICITATION

Siobhan Warrington

ABSTRACT

This chapter introduces the approaches and methods employed in a four-country research project that resulted in the 2017 report The People in the Pictures: Vital perspectives on Save the Children's image making. *It presents and explores the ethical issues that emerged throughout the process of the research, particularly in relation to photo elicitation – the use of images (still and moving) within both interviews and focus groups. Interviews and focus groups took place in the UK, Jordan, Bangladesh, and Niger with a total of 202 research participants. The research involved sharing Save the Children content (fund-raising materials, published reports, online news features, TV adverts, and short films) with research participants Research participants included those featured in some of these visual communication materials (referred to as contributors), and other individuals within their communities (referred to as non-contributors). The following principles and decisions informed the research design: safe and ethical practice; inclusive, engaging and accessible approaches; the participation of children; prioritising first-hand accounts; no photography or filming; and the preparation of location- and language-specific resources for each interview and focus group. The main ethical issues to emerge during the design of the research related to predicting (and responding) to any potential negative impacts of the research on participants, particularly contributors, but also children. The researchers also experienced some unexpected ethical encounters, including visual materials causing some concern or distress. Additionally,*

Ethics and Integrity in Visual Research Methods:
Advances in Research Ethics and Integrity, Volume 5, 43–63

ISSN: 2398-6018/doi:10.1108/S2398-601820200000005007

assuring research participants' anonymity led to the necessity of extra care when publishing the report and the use of images within that.

Keywords: Photo elicitation; research ethics; visual methods; contributors; NGO communications; reflective practice

This chapter introduces the four-country research project that resulted in the 2017 report *The People in the Pictures: Vital perspectives on Save the Children's image making* and explores the ethical issues surrounding the use of images (still and moving) within the research. Despite the extensive research and debates on visual representations of global poverty, the voices of the people featured in those images have been notably absent. *The People in the Pictures* research addressed that gap by exploring how those featured, and their communities, experience and perceive the process of image making, and how they respond to their portrayal in Save the Children's visual communications.

The research involved sharing a selection of Save the Children's visual communication materials (also referred to as Save the Children content) with a total of 202 research participants including contributors, non-contributors, and a small number of Save the Children staff. The term 'contributor' refers to those who have contributed their image and story to Save the Children for fundraising, campaigning, and/or awareness-raising purposes. It is a deliberate alternative to 'subject' to better acknowledge their active role in, and contribution to, image making. 'Non-contributor' refers to research participants who are members of contributors' communities[1] or similar communities, but who have not contributed their image or story to Save the Children's visual communications.

I am adopting a standard definition of ethics and ethical issues to mean moral principles and the dilemmas encountered over what course of action is right or wrong. Within this chapter, I will explore the considerations and actions taken by myself and other researchers to avoid causing any discomfort or harm to research participants throughout their participation in the research. This includes the deliberate research design decisions to implement research that was safe, inclusive, accessible and engaging, and the additional care required when using images in the published report in order to adhere to our promise of anonymity for all research participants.

The chapter examines the ethical encounters surrounding the use of photo elicitation within the research. These include our preventative actions as well as ethical issues we did not anticipate. The chapter presents these specific ethical encounters related to the use of images within the wider context of the research and its approach.

In order to widen the knowledge and perspectives upon which this chapter is based, I have included the experiences of two other researchers who also used photo elicitation in recent research initiatives on NGO images. Dr E. Ademolu's (2018) doctoral thesis, *Rethinking Audiences: Visual Representations of Africa and the Nigerian Diaspora*, was based, in part, on a series of focus groups and interviews with Nigerian diaspora to explore their response to a selection of NGO images. *A Study of Visual Communication in Six African Countries* by D. Girling (2018) and RADIAid[2] presents the responses of young people from Ghana, Zambia, Malawi, Ethiopia, Uganda, and South Africa, to a series of NGO images. Ademolu and Girling both shared with me their reflections on the ethical issues relating to their use of images within their respective research projects. This chapter also benefits from input by A. Gormley who carried out most of the interviews and focus groups for *The People in the Pictures* research. She shared her additional recollections of specific ethical encounters and those have been included within this chapter.[3]

INTRODUCING THE RESEARCH

> Debates about representations of global poverty and images of suffering have been going on for many years, yet the voices of the people featured in the images – the contributors – have been notably absent. There has been a tendency to consider and judge the image alone, rather than recognise it as the result of a process involving multiple stakeholders, one of whom is the person in the image. (Warrington & Crombie, 2017, p. vii)

Save the Children UK (SCUK) commissioned *The People in the Pictures* in order to learn more about its own image making from the perspective of those featured in its visual communications. The research aimed to 'influence image making and use across the Save the Children network … (and) contribute to the debates on representation and the image-making process within the wider INGO sector …' (Warrington & Crombie, 2017, p. 3). It was rooted in Save the Children's long-standing commitment to reviewing its policies and practice relating to image gathering and use.[4]

The People in the Pictures built on the previous unpublished, overseas component of the *Depicting Injustice* research commissioned by SCUK in 2010 (Miskelly & Warrington). *Depicting Injustice* engaged with contributors and their communities in the Occupied Palestinian Territories, India, Kenya, and Afghanistan to explore their views on visual representation and their experiences of the process of photography. *Depicting Injustice* also included participatory photography as a research method, supporting adult female research participants in three of the research locations to take their own photographs and caption those in order to further understand their preferences for the visual representation of themselves, their children, and their neighbourhoods.[5]

The *People in the Pictures* research sought contributor and non-contributor responses to Save the Children's finished visual communications, including film (moving images) in addition to printed or online communications materials containing still images. It was an exploratory and qualitative research project, commissioned by SCUK but designed and led by external researchers.[6]

By placing contributors and their communities at the centre of learning about image making, the research embodies the shift in practice and debates it aims to provoke. It does this by: recognising the contributor as a key stakeholder in NGO image making; emphasising the importance of contributors' perspectives to debates on (and knowledge of) NGO image making and representations of global poverty; and recognising the image-making process as a site of enquiry.

The research was structured around three areas relating to image making from the perspective of both contributors and non-contributors: motivations, process, and portrayal. The research aimed to explore people's motivations to agree (or not) to Save the Children filming or photographing them or their children; their experiences and perceptions of the image-making process; and responses to their portrayal in resulting Save the Children communications, including the imagined and real consequences – positive and negative – of this portrayal.

Primary data collection took place in several sites across four countries between November 2014 and July 2015: UK (East Anglia, London, Kent, Wales, and Yorkshire); Jordan (Amman and Zaatari Refugee Camp); Bangladesh (Dhaka and rural Habiganj); and Niger (Niamey and rural Zinder and Tessaoua). At the time of the research, none of these were the location of sudden emergencies.

There were a total of 202 research participants: 53 contributors[7]; 130 non-contributors; and 19 staff members. They included children (aged 7–17 years), young adults (aged 18–21 years), and adult parents.[8] Across the research, there were 39 interviews (with contributors and staff) and 21 focus groups (mostly with non-contributors). The interviews and focus groups were semi-structured following a set of questions and activities framed by the overall research questions. The questions, activities, and visual content used for each different interview and focus group were adapted according to location, age, and, at times, gender.

In 2018, SCUK established its Research and Evaluation Ethics Committee (REEC), made up of experienced, and independent, researchers. All research undertaken and/or commissioned by SCUK is now reviewed by the REEC before any primary research can take place. At the time of conducting *The People in the Pictures* research, there was no formal ethical review process in place for SCUK research initiatives. During the inception phase of the research, the lead researcher produced a detailed Research Outline for review by SCUK which included a section on ethical considerations, with a particular focus on the participation of children in the research and plans for informed consent. The plans for including children in the research were reviewed and approved by SCUK's child safeguarding manager. All UK-based researchers who were due to come into contact with children throughout the research underwent Disclosure and Barring Service checks. Interviews and focus groups with children were carried out in accordance with Save the Children's code of conduct and child safe-guarding policy and these also formed part of all researchers' and interpreters' contracts.

Ethical considerations must, however, go beyond a series of reviews and checks at the beginning of a research project, and so researchers were required to think critically and ethically throughout the research process. All those involved in conducting (and supporting) interviews and focus groups (researchers, assistants, and interpreters) for *The People in the Pictures* were encouraged to reflect upon and critique each interview or focus group. Researchers adopted a reflexive approach and their experiences, and reflections of interviews and focus groups were documented and discussed. Successes and challenges within the research experience were openly discussed with a view to improve successive interviews and focus groups. This critical reflection allowed for unexpected issues to be considered and resolved, and it was especially important for a piece of research that at the time was relatively novel in its approach. There were no similar initiatives for the researchers to refer to; no lessons learnt shared by others to inform the design and approach of *The People in the Pictures.*

MODELLING THE PRINCIPLES OF PARTICIPATION AND PROTECTION

For a research project addressing issues of responsible practice and representation, it was imperative that the research design be based on ethical considerations and that the principles of protection and participation were modelled throughout the research. An awareness of the unequal social-political system in which the research takes place – a system within which the researchers had considerably more power and privilege than the research participants – also created a necessity to design the research in a way which was safe and meaningful for research participants.

Research design was informed by the following commitments:

- To be safe and inclusive, especially regarding the participation of children.
- To be engaging and accessible for everyone.
- To prioritise first-hand accounts.
- To be location- and language-specific.
- To use images as a tool to generate discussion, not for comparable data collection.
- To not film or photograph any parts of the research.

Safe and Inclusive, Especially for Children

Methods and approaches were designed to create a safe, sensitive, and inclusive research environment for all research participants, with participation based on their informed consent in advance of the interview or focus group. With so much of Save the Children's content featuring children, it was essential to involve children in the research. We developed specific child-friendly approaches for interviews with child contributors and focus groups with both younger children (aged 7–12 years) and older children (aged 13–17 years). Inevitably these tended to be considerably shorter in duration than interviews and focus groups with adult research participants.

All interactions with research participants were managed by Save the Children staff or staff of partner organisations who had existing relationships with participants. Many of the focus groups with children took place at Children's Centres run by Save the Children, with staff from those centres present throughout. For these staff, the children's well-being took priority over the research. If any children had felt upset by any of the images used in the research (none did as far as we know), we were confident that the support structures were in place to respond immediately and appropriately.

In each research location, we ran separate focus groups with older and younger children in which many questions (and their responses) were game-based or involved physical movement. Even for research using images (which might be considered more accessible than word-based interviews or focus groups) physical movement and games were still important, especially for the focus groups with younger children.

Engaging and Accessible for Everyone

The principles of accessibility and engagement are also critical when it comes to designing interviews and focus groups with adults. The researcher must balance seeking answers to the research questions with a consideration of participant experience, ensuring that the process is safe, meaningful, accessible, and engaging for those who are sharing their time, experiences, and perspectives with the researchers. For example, within the adult focus groups, coloured cards were provided to enable all participants to signal their individual and personal response to questions posed by the researcher (a green card for yes and a pink card for no). This interactive approach helped to avoid the prevalence of group think within the focus group.

The use of interpreters was required for all contributor interviews and focus groups, apart from those in the UK. Most contributors interviewed were women and for that reason only female interpreters were recruited to support the interviews and focus groups in Jordan, Bangladesh, and Niger.

Prioritisation of First-hand Accounts

The research was underpinned by a belief in the importance and value of personal, subjective and verbal accounts of those taking part, and so all interviews and focus groups were audio recorded, transcribed, and translated. Given the absence of the voices and perspectives of contributors and their communities in debates on NGO image making to date, it was important to use verbatim accounts within the report itself.

Location and Language Specific

In total, there were 51 different pieces of Save the Children content used across the research, the majority of which had been produced between 2012 and 2014. The majority of those used within the research had involved a professional image maker, and in the case of Jordan, Bangladesh, and Niger, very often an image maker from overseas. Examples of those used in focus groups are provided in the final report (Warrington & Crombie, 2017, p. 12).[9] As far as possible we aimed to include a good range of the different types of visual communications materials produced and used by the Save the Children: TV adverts, on-line films, research reports, fundraising materials, and news features. We also included at least one example produced by the Save the Children country office where the research was taking place.

Many of these visual communication materials had to be translated from English into languages understood by the research participants before they could be used within the research. Films were dubbed or sub-titled; and captions and other text contained on printed/online materials were translated. The researchers acknowledged that some research participants would be non-literate, but this advance translation would enable interpreters to accurately share these resources verbally during the research.

Each interview and focus group had its own bespoke set of visual resources and accompanying questions; meaning the researchers created around 40 distinct sets of visual resources and accompanying interview and focus group guides.[10] Preparing these distinct resource sets for each interview and focus group was time-consuming but considered essential to ensure research methods were inclusive, accessible, and engaging. It was also important to create sets that were suitable and appropriate for focus groups with children, especially younger children.

Using Images (Still and Moving) as a Tool to Generate Discussion

Throughout the research, images were used as a tool to generate discussion – to encourage research participants to share their perspectives on portrayal and visual representation, as well as their experiences and views on the image-making process. The research deliberately did *not* attempt to generate comparable data on specific images. The research aimed to extend the debate on representation beyond analysis and debate over particular images, to include consideration of the contributor experience of image making. The research consciously adopted a framing of image making that in terms of enquiry, prioritised the *process* of photography and filming over the image as the *object* of focus. On reflection, this framing shares some aspects of the rationale proposed by Azoulay (2015) of understanding photography as an event, with participants other than the photographer and outcomes other than the photograph. Fairey (2015) provides a useful summary of Azoulay's revised notion of photography:

> an ontological investigation of photography that moves beyond a focus on the technology of the camera or the photograph alone … photography is reconceived as an event, ongoing and never-ending with multiple participants and under the sole control of no one. (pp. 22–23)

No Filming or Photography during Research

It was important to both Save the Children and the researchers that participants clearly understood that the research activity was distinct from the image making to which they had contributed. For that reason, a decision was made to not take any photographs or to film during the research, as to do so could have compromised that distinction. This decision was a transition from the previous *Depicting Injustice* research within which we had taken photographs of the contributors we interviewed together with the finished content in which they were featured. Some of those images are particularly powerful, but it was decided for *The People in the Pictures* research that it was important to sacrifice these potential research outputs for an approach that we felt was more ethical.

Limitations and Biases

It is important to acknowledge that, despite the research design decisions outlined above, limitations and biases remained. Of greatest significance was the lack of a shared language between researchers and research participants. In Jordan, Bangladesh and some parts of Niger, the researchers relied on the assistance of interpreters to conduct interviews and run focus groups. Although we worked with experienced interpreters, and all sessions were recorded, transcribed, and translated into English, the use of interpreters limits both the opportunities for research participants to fully express themselves to the researcher and the researcher's understanding of the research encounter.

All messaging accompanying the visual communications used in the research was translated and researchers carefully explained the purpose of these communication materials to research participants. We cannot, and should not, however, assume that this context and meaning was always clear to all participants. Contributors' realities and contexts were often different to those of the audience for whom the visual communications were intended, and as such, some of the messaging and overall intention of the communications may not have been fully relatable to some participants, despite careful translations and explanations.[11]

There was an unavoidable bias within the sample of research participants, based on their existing relationship with Save the Children. By necessity, all contributors interviewed were those who Save the Children were able to contact; such individuals were likely to be those who had a recent and positive experience with Save the Children, and who were therefore perhaps less likely to be critical of Save the Children in interviews. The researchers also acknowledged that the same reasons that influence contributors' decision to contribute their image and story to Save the Children will have influenced decisions concerning participation in the research. This decision is not neutral when the organisation requesting your participation (in either communications or research) is the one that you are benefitting from or have benefitted from in some way. This is also acknowledged by Siobhan McGee (2005) in her review of the 1989 *Code of Conduct: Images and Messages related to the Third World:* 'The relative power of beneficiaries is low … even if permission is granted it may not be fully voluntary if their livelihood is associated with the requesting NGO' (p. 6).

While the researchers were external to Save the Children, for participants the interview or focus group will have been perceived as a Save the Children activity (interviews and focus groups were arranged by Save the Children staff and often located in a Save the Children space). Within this context, many research participants (some of whom at the time were benefiting from Save the Children supported services or programmes) would be unlikely to openly criticise Save the Children's content or image-making processes.[12]

In response to this bias, the researchers explicitly assured participants that their responses to interview or focus group questions would in no way impact on their existing relationship with Save the Children. All participants were assured anonymity in the published report and so no links were made between research participants and the Save the Children content they contributed to. Within the report, you will not find any extracts from contributor interviews alongside images from the Save the Children content within which they appear in and are visibly identifiable. Instead, most of the images presented in the published report are from the content used within focus groups, and interviews, that does not feature any of the contributors engaged in the research.

THE USE OF PHOTO ELICITATION: UNDERSTANDING THE ETHICAL RISKS

Photo elicitation refers to the method of using images within interviews or focus groups as a stimulus for response. It has been, and continues to be, used across various disciplines and includes different approaches, including research participants' taking their own photographs and talking about those, as well as viewing and discussing photographs or films made by others.

Given the aims of the research, *The People in the Pictures* relied heavily on photo elicitation as a method, using both still and moving images throughout all interviews and focus groups. The creation of bespoke sets of visual resources for each interview and focus group is outlined earlier in this chapter. For each country research location, we developed a set of what we referred to as 'generic content' this was a set of visual resources that would be used (at least in part) across all of the focus groups and interviews within that country. Interviews and focus groups always started with a set of straightforward questions or activities designed to explore participants' current relationship with visual media. All interviews with contributors involved viewing and discussing one or more pieces of Save the Children content that featured them (or their child) and this always took place after the initial discussion on their relationship with visual media. Within contributor interviews, their memories and experiences of the process of being interviewed and filmed or photographed followed on from this, with the visual content serving as a reminder of that particular experience. We also shared a selection of the generic set of content during contributor interviews in order to explore and capture their views on the visual representations of others. Focus groups followed a similar structure.

The widely cited article by Harper (2002) *Talking about pictures: A case for photo elicitation* makes a strong case for the value of photo elicitation as a research method that 'enlarges the possibilities of conventional empirical research' (p. 13). Harper argues that, 'photo-elicitation produces a different kind of information. Photo-elicitation evokes information, feelings and memories that are due to the photograph's particular form of representation' (p. 13).

These reasons cited by Harper for the value of photo elicitation – the qualitative power of the method – would suggest that it carries with it an additional ethical burden. If photo elicitation can evoke, as Harper claims, feelings and memories – 'deeper elements of human consciousness' (p. 13) than words alone, then surely it also has the potential to trigger negative emotional responses in research participants which may create distress.[13]

The use of images within *The People in the Pictures* research presented researchers with the practical challenge of preparing bespoke sets of user-friendly visual resources, but it also created a particular set of ethical concerns to consider and resolve. The over-arching ethical concern relating to the use of still and moving images within an interview or focus group is consideration of whether those images will cause any distress or offence to research participants.

Multiple Responses and Difficult Predictions

In *On Photography*, Sontag (1977) states 'Photographs, which cannot themselves explain anything, are inexhaustible invitations to deduction, speculation and fantasy' (p. 23). I take her statement to mean that images are a vessel for multiple subjectivities. We can appreciate that certain images, by their nature, might cause distress, and as researchers, we can make decisions about whether to use those in our research, and whether we need to use them cautiously. However, we can never entirely predict how a research participant will react and respond to an image, as this can be heavily influenced by the research participant's unique biography, which the researcher is unlikely to know in its entirety.

The experience of *The People in the Pictures* research demonstrated the difficulty of predicting participants' reactions to images, in terms of the plurality of responses documented across interviews and focus groups. Within one family in Jordan, there were several conflicting responses, positive and negative, to the images of family members that had been published in a Save the Children report. In some interviews, contributors clearly experienced a sense of delight at seeing themselves or their children in the visual communications we shared, and in the first interview undertaken for this research, the researcher walked into the contributor's flat to find some of the images taken by Save the Children of her and her children framed and hung on her walls. In contrast, a contributor in Bangladesh explained how she felt the photograph returned to her, as part of the research, represented her vulnerability; it was not something to feel positive about.

> If I had a good life then I would feel good, but in reality … I am in this profession (sex work). Now, nothing can be done if I feel good or if I feel bad. If I were in a good position then I would be able to say that this is a good photograph …. Looking at this photo, I feel like I am in a vulnerable position. I am vulnerable, that's why the photo was taken. (Habiba, adult female, contributor, Bangladesh)

This experience of different responses demonstrates the impossibility of researchers fully anticipating how participants will respond to images and therefore the difficulty of mitigating all ethical risks within the research. While the impossibility of predicting research participants' responses is not specific to the use of photo-elicitation, as Sontag suggests images by their nature are open to a greater degree of interpretation than words alone.

For all researchers, there are ethical issues which can be predicted and prepared for; and there are those we do not predict, that are unexpected and to which we may or may not have an appropriate response at the moment we encounter them.

In other words, there are ethical concerns and dilemmas which are considered and responded to by the researcher in advance of the fieldwork, those that constitute the ethical preparedness of the researcher; and there are the unexpected ethical dilemmas and problems encountered that test researchers' ethical responses. Reflecting on these, in the context of *The People in the Pictures* research, and the research of others, provides lessons for both future research practice as well as insights that are relevant to the research topic itself – NGO image making.

The Relationship between Research Participants and those Featured in the Image

Ademolu's reflections about the potential of NGO images used within his research to cause distress, relate to the relationship between his research participants and those images. He referred to having to consider the added racialised component of conducting research with communities of Black African heritage who have their origins in the places and with the communities that are portrayed in NGO communications. He explained, how, he anticipated that for some participants involved in his research,

> they might have negative emotional responses to photos – depicting (and which allude to) Black African trauma and suffering – conjuring painful first-hand or vicarious memories, that were perhaps archived away and suppressed. I also considered the ethics around photos causing embarrassment and indignation. (E. Ademolu, personal communication, January 2019)

Ademolu's account above describes the potential risks associated with using images that feature people with whom research participants share aspects of their identity. Does the closer the relationship or connection between the image and the person being interviewed, increase the chances of the image evoking strong feelings; do the ethical risks increase? If so, we could consider *The People in the Pictures* research at the sharp end of that spectrum given that we were returning images to the people featured in those images, and members of their (or similar) communities?

Most contributors (in Jordan, Bangladesh, and Niger) were glad to see and receive the images of themselves or their children for the first time, with some requesting to watch the films featuring themselves or their child several times during the interviews.[14] For some contributors, however, their feelings about seeing themselves (or their children) in SCUK content were clearly more complex and layered. They were appreciative of having the images of themselves or their child returned, however, in a few cases the images were also a visual reminder of their child's serious illness, or their own poverty. The emotional response to images of poverty or ill health is likely to be more significant when you are looking at a picture of yourself or your child. Indeed asking how people feel about their representation, who are being represented due to their poverty, is not a neutral question, as the interview extract on the previous page with Habiba illustrates. The ethical issues arise not just from the choice of images used, but also the identity and immediate circumstances of the research participants. In the case of *The People in the Pictures*, many could be considered relatively vulnerable or lacking in power; they included: children, refugees living in a camp, those living in poverty, and parents of previously malnourished children who received health care by Save the Children or its partners. Sensitivity, care, and respect were essential qualities for all those designing and undertaking interviews and focus groups within the research.

ETHICAL PREDICTIONS

During the identification and selection process of the visual communication materials to share and discuss in each research site, a number of predictions emerged relating to the potential for the content to cause offence or distress to research participants. The ensuing adaptations to the research approach or methods in particular contexts, and or with particular groups of research participants can be referred to as ethical preparedness.[15]

Designing Interviews and Focus Groups with Children: Care with Content and Negative Messaging

Within *The People in the Pictures* research, the well-being of child research participants always took priority when designing interviews and focus groups and selecting which visual content to use within those.

All of SCUK's film content is categorised as U, however, for the researchers this did not necessarily mean it was automatically appropriate for sharing with child research participants. One of the films we had selected to show to children in the UK focus groups was a short film about Save the Children's Families and Schools Together Programme (FAST). Upon viewing the film, the researchers decided to show only the first 23 seconds of the film; the messaging in the rest of the film associates childhood poverty with the possibility of lower educational attainment. It did not feel appropriate – in fact, it felt potentially harmful – to show the rest of the film to children, some of whom would be considered to be living in poverty. In addition, the focus groups were taking place at the children's school, and the messaging of the film – directed at individual donors and policymakers – would be somewhat in contradiction to the mission and ambition of the school which seeks to support all children to achieve their full potential.

A similar dilemma was predicted for a contributor interview in Jordan, for which the researchers felt part of the short film featuring a child had the potential to be mildly distressing to that child – due to its general message about child poverty.[16] Given that the interview was being carried out with the child's parents, with some input from the child, the researchers decided to raise their concerns with the parents prior to the viewing. Both parents felt comfortable with their child being present while the film was shown during the interview and neither parents nor the child appeared to find the film (or its messaging) distressing in any way.

Across all research sites, we made efforts to find content which we believed carried minimal risks for young children in terms of the content having any negative emotional impact. In Bangladesh, it was known that all children participating were visitors to Save the Children's centre for street children, and who were therefore all likely to have experienced some form of neglect or abuse. With this knowledge, the researcher took additional care in relation to the selection of images shown in the focus groups with both younger and older children.

The criteria for selecting visual communications for use in interviews and focus groups with children included:

- content featuring children;
- content including children's voices;
- content that was deemed engaging for children; and
- content selection to include children who were recognisably happy or having fun as well as content featuring children who could be seen to be in need.

The Protection versus Participation Dilemma

The content relating to one contributor in Bangladesh included a poster intended for use on the London underground. The poster featured a photograph of her and her crying baby, who was sick at the time, and the message 'He's so sick, he could die before you've read your paper'. While the messaging is obviously, in part, based on evidence of infant mortality rates, it is nevertheless designed to shock the audience. I spent a long time considering whether to show the contributor this poster as part of her interview; thinking through how to balance her rights to see the content she had contributed to and to participate in a conversation about that, versus, our responsibility to protect her from become distressed as a result of participating in the research.

Ultimately, the researchers decided it was important for the contributor to have the option of seeing the content she featured in. We decided to manage the risk of causing her distress with a careful introduction to the content and providing her with the option of not looking at it. The contributor chose to view the content and appeared to not be distressed by the image or messaging. However, the dilemma is itself an important outcome of the research – a reminder that the messaging that accompanies images is often conceived in a different time and space than the image-making process, without consultation with the individual being represented.

The ethical issues experienced by Ademolu and Girling in their research also related to the potential of the images used within interviews or focus groups to cause distress or offence. Girling describes a similar strategy to that described above, in that at the beginning of each focus group for the research undertaken with RADIAid, the facilitator told participants that they will be shown 10 images from real NGO campaigns in Europe and Australia including two images of child soldiers and an image of a malnourished child. Participants were given the option to withdraw from the group if they felt they could be disturbed by these images; while some participants were clearly saddened by the images, none were visibly distressed, and no one selected to withdraw from the focus group (D. Girling, personal communication, 15 January 2019).

Ademolu (2018), recognising that 'presenting and discussing images of human suffering, impoverishment and environmental catastrophe could conjure negative, emotional/psychological responses of distress and/or embarrassment among group discussants' took the additional step of preparing a list of local support services should any research participants feel the need for support following their participation in the research (p. 87). As far as Ademolu is aware, this was not used by any research participants.

Ademolu also explained that he made a deliberate decision to not use the images used by lesser-known poverty charities, images which he felt were 'grotesque almost pornographic portrayals of children and families who appear as nothing more than curiosities of the flesh... that are seemingly worthy of benign (largely white) Western contemplation'. He further explained,

> If I had gone down that route, I would not only be breaching all ethical considerations that advise against harming others but personally, I felt I would be doing these communities a disservice by agitating or 'stick poking' their emotional wellbeing. (E. Ademolu, personal communication, 9 January 2019)

UNEXPECTED ETHICAL ENCOUNTERS

Messaging as a Trigger to Recollecting Personal Trauma

The first focus group conducted as part of this research, took place in South East England, and was undertaken as a pilot. As with the majority of focus groups, participants were not contributors themselves, but nevertheless had a relationship with Save the Children through its programming or projects. This first focus group in the UK involved several parents of children at a school which participated in SCUK's FAST programme.

During this focus group, the set of visual communications selected and used included a report with references to stillbirth and miscarriages, as part of SCUK's 2011 No Child Born to Die campaign to combat infant mortality. One of the focus group participants had a quick look at this report and quickly turned away stating that she could no longer look at it, having experienced a stillbirth herself. The researchers paused the focus group in order to remove the report, apologise, and check whether the participant wanted to continue participating in the focus group; she did.

This report was subsequently removed from the research due to the risk of causing distress to other research participants who may have experienced a stillbirth or miscarriage themselves. (Statistically, the likelihood of participants in any adult female focus group having experienced either of these is significant.) The experience provided researchers with an early indication of the potential for images used within the research to cause distress to research participants.

On reflection, were we as researchers guilty of failing to appreciate that communication materials featuring women from overseas could relate to the experiences of female research participants in the UK? Did we just not look closely enough at the messaging, or did we simply assume this report did not need checking so carefully because it was not about UK poverty?

Previously Unseen Images Causing Concern

Researcher A. Gormley recalled a difficult moment during the final contributor interview she carried out in the UK. This interview was with Wendy,[17] a mother who had been photographed and interviewed in relation to SCUK's poverty programming. The day that Anna went to do the interview, was the day Wendy's husband had temporarily moved out of their home as a result of his post-traumatic stress disorder (something that was not known to either the researcher or Wendy at the time of arranging the interview). As with all interviews, the researcher shared and talked about the Save the Children content Wendy had featured in. At the point that Gormley showed Wendy the final piece of content, Wendy expressed surprise and some distress. This was a piece of content she had not been made aware of by SCUK, and what particularly concerned her was that it was available online. She was worried about how her husband would react if he came across it. Immediately after the interview Anna called the staff member at Save the Children who had arranged the interview to report what had happened. Nevertheless, as a researcher, Gormley felt that unintentionally, she had left Wendy feeling more distressed than she had been before the interview and wondered whether anything could have been done to avoid this.

This experience highlights the need for additional consent when organisations decide to use images in ways beyond what was originally envisaged and communicated to the contributor. During the time the research was being carried out, checking consent for re-use did become standard practice in the UK (see p. 43 of *The People in the Pictures*).

Returning Image Leads to Contributor withdrawing Consent

The focus group with women in Zaatari camp, Jordan involved several women who had contributed to a Save the Children photography initiative for a BBC news feature in 2013. During the focus group, the response to the images featuring some of these women demonstrated what became an important research finding: the potential for contributors to misunderstand the purpose of Save the Children's image making.

While looking at the images, the focus group participants were asked if they could recall why the images had been taken. The transcript from this focus group reveals that there were multiple understandings of the purpose of this photoshoot among the women and there was a lack of clarity for some about why they had been asked to be photographed. It is not appropriate to simply assume that this was a result of a poor explanation by Save the Children at the time the photographs had been taken; rather we need to appreciate that regardless of how it was explained, not all women will have understood the explanation provided in the same way. It was also clear that for some there was an existing concern about photography that stemmed from the presence of a UNICEF and Noor[18] photography initiative which had resulted in some large-scale portraits of female residents of Zataari camp being pasted onto the walls of the camp.

Researcher A. Gormley recalled how for a small number of women, given their lack of clarity over the ultimate use of the images, reviewing these images caused some concern and worry. A few of the women were concerned with how these images may be read by others and were particularly worried that their husbands may end up seeing them. One woman clearly stated that she no longer wanted her image to be used by Save the Children. This request was promptly actioned by SCUK who removed her image from their image bank.

It is regretful that returning the images to women contributors as part of the research caused a small number of them additional concern. However, this act of returning the images allowed for one of the contributors to take control over her image and to make a more informed decision to withdraw consent upon seeing the finished content. This instance and the one in the UK described above demonstrate the importance of returning finished content to contributors who may change their mind about the use of their image or story for any number of reasons.

These two encounters both relate to the real or imagined responses of people known to the contributor seeing the visual communications they feature in. The research suggests that this was more likely to be a concern for contributors in the UK and Jordan who had greater, regular access to visual media than research participants in Bangladesh and Niger (particularly rural Bangladesh and Niger).

CONCLUSIONS

The People in the Pictures: Vital perspectives on Save the Children's image making is a research report published by an international NGO based in the UK about how contributors and their communities based, mainly, in the global South experience image making. It is authored by two white women. As such, to some extent it replicates the inequalities that frame the international development and humanitarian aid system.

Nevertheless, the report, and the way the research was carried out, did aim to disrupt existing debates by prioritising the views and experiences of contributors and their communities, over a focus on the image, the image maker and how images are received by audiences in the global North. The research provides unique qualitative evidence from the perspectives of contributors to justify the need for responsible NGO image making.

The decision to interview Save the Children staff in each of the countries where the research was carried out, and to actively seek their feedback and input into various drafts of the report, ensured professional expertise from the global South informed the research and resulting report. Including the UK as a research location, alongside Bangladesh, Niger, and Jordan also served to widen the debate. In particular, the research demonstrated different experiences of image making for contributors in the UK, as compared to other research locations, particularly in relation to Save the Children's communication with them before, during, and after photography or filming.

For a research project that was concerned with the issues of representation, communication, agency, accountability, process and power, research design decisions were not merely methodological, they were imbued with an ethical and moral requirement to be aware of the unequal power relationships within the research context, and to wherever possible avoid replicating or reinforcing those within the research itself. Specifically, a research project that explored informed consent had to ensure its own informed consent procedures were meaningful for research participants.

Reflecting upon the ethical issues relating to the use of photo elicitation with *The People in the Pictures* research serves to highlight specific lessons for both responsible research practice and responsible image making.

Within *The People in the Pictures* and also Ademolu's and Girling's research, research participants were all alerted to the types of images that would be shared with them throughout the interview or focus group. This advanced notification supported their informed decision to participate in the research or not. This level of transparency is important when using photo elicitation responsibly within interviews or focus groups.

The experiences from *The People in the Pictures* research, and Ademolu's research, suggest that there are layers of risk relating to research participants' connection or relationship with the images used in the research. The research suggests that the ethical stakes may be higher the closer the relationship of the research participant to the image being looked at and discussed, be that a relationship based on (a) aspects of one's identity or (b) shared experience with those featured or the issues being illustrated in the image. As part of their preparation, researchers planning to use photo elicitation could ask themselves the following questions to help surface and predict whether the images used may be more likely to result in a heightened emotional response by research participants:

- Does the research participant share some aspects of the background of those featured in the image(s)?
- Could the research participant(s) have had similar experiences to those portrayed in the images?
- Are research participants featured in the images being discussed?
- Have the research participants seen the images before?
- Did research participants take the photos themselves?

An examination of the ethical issues emerging from the use of images within the research also reveals a series of lessons for more responsible image-making practice.

First, NGOs must always provide contributors and potential contributors with clear explanations of the purpose of the image making and the intended audiences. These explanations must always be given in a language that the contributor understands.[19] We must, however, acknowledge that while a clear explanation is critical, it cannot guarantee that all contributors will always come away with the understanding that those involved in image making intended – as the example outlined above with women in Jordan demonstrates. It is up to NGOs to continually strive to find effective ways to communicate explanations of the purpose and use of images with potential contributors.

Second, we must acknowledge that images and the messages accompanying those images are often created in different spaces and at different times. The text or message that accompanies an image, that translates the image into a purposeful piece of NGO communications, is generally developed at a time and space away from the contributor. At the time of gathering informed consent, image makers may be unaware of the exact messaging that will be used alongside a contributor's image, and explanations about purpose and use may therefore be general. It is, after all, difficult for those present at the time of filming or photography to be confident about the final messaging that a fundraising team based elsewhere will decide to attach to an image. What is less difficult, however, is for those involved in content production to consider how the contributor might feel seeing that content – to employ a degree of empathy when representing others in public communication materials.

Third, we must consider the NGO's responsibility to gather consent for re-use. A typical informed consent process should make it clear to the contributor that they are consenting to the NGO to use their images, in a variety of ways, for a certain length of time – usually either for three or five years, thereby avoiding the need to gather consent for re-use.[20] However, NGOs should always consider the age of an image, and whether it is still appropriate to use it, in relation to the potential for the contributor and/or their circumstances to have significantly changed since the image was taken.

In the past, concerns about contributor satisfaction with final content and contributor permission for re-use of images may have been limited to contexts in which contributors are recognised to be part of the audience for the intended communications, for example, UK contributors who have been filmed or interviewed for a UK fundraising campaign. Content for a UK audience featuring contributors from the global South may have felt less problematic in relation to taking the necessary time and actions to ensure those contributors are comfortable with their representation. Afterall, it would be unlikely for those contributors to end up viewing that content. The idea of a distinct audience, however, that is, separate from the contributor and their community, is increasingly out-dated. Ever-increasing smart-phone ownership, internet connectivity, and the use of social media mean that the potential audiences for NGO communications can include contributors' communities and family members. This changed context provides a greater urgency for those gathering images for NGO communication purposes to ensure that informed consent procedures are effective and meaningful for the people whose images and stories become part of their communications. Contributors' consent must be based on an understanding that anyone, including their family members, community and employers could potentially see their image or read their story.

Most of the contributors from Jordan, Bangladesh, and Niger who participated in the research had not seen the finished Save the Children content featuring themselves or their children, prior to the interview or focus group. The research highlighted the value and importance of returning images and recommends that wherever possible this should be part of NGO image making. Follow-up with contributors and returning content and/or images is not just a way of checking consent for future use of images, it can assuage any concerns relating to uncertainty over the use of images, it can help manage expectations, and it provides an opportunity to check contributors' well-being. Most importantly, it demonstrates respect towards contributors and the value of their contribution. In the words of one SCUK staff member,

> It's about keeping promises… and making sure they feel safe and looked after. It's about consistency – that they always get copies of things to keep … it's just treating them as part of a team, in the same way you would if somebody had done a fantastic piece of work for you.

FINAL NOTE

Jess Crombie who commissioned the research was determined that the research fulfil its aims of generating debate (and change) across the sector of UK organisations working in international development. On the day *The People in the Pictures* report was launched, Jess brought together a group of communications and fundraising directors from across the sector to discuss the implications of the research. The group has continued to meet on a quarterly basis since June 2017 and in 2018 become a Bond working group with Jess Crombie and Tamsin Maunder Betti serving as chairs.[21] The Bond People in the Pictures group serves to provide sector leadership on ethical approaches to gathering and using images. It is also a space for peer discussion and support on best practice approaches. Siobhan Warrington was commissioned by the group to develop a new set of ethical content guidelines for Bond members. *Putting the people in the pictures first: Ethical guidelines for the collection and use of content (images and stories)* were incorporated into Bond's charter in November 2019.

NOTES

1. The term "community" is used here in the broadest sense to refer to those living in the same geographic area, but also to those who may be connected to others through shared identity or interest.
2. https://www.radiaid.com/ An annual campaign to challenge perceptions around issues of poverty and development.
3. Anna Gormley is a tutor and ESRC-funded PhD candidate at JOMEC (the school of journalism, media and culture), at Cardiff University. She also works as a producer in social documentary photography. Her practical and theoretical work has explored notions of collaboration and the re-imagining of the production of photography, most recently in a humanitarian context.
4. For example, *Focus on Images: The Save the Children Fund Image Guidelines (1991)*, *Interviewing children—a guide for journalists and others* 3rd edition (2003), *Depicting Injustice* research (2010).
5. Within some literature, the use of participatory photography (or photo voice) within research is referred to as auto-driven photo elicitation, for example, Ford et al. (2017) Auto-driven Photo Elicitation Interviews in Research with Children: Ethical and Practical Considerations.
6. It was commissioned by Jess Crombie in 2014 who was, at the time, Director of Creative Content at Save the Children UK. Jess provided support with research design, analysis and research communication. Logistical support was provided by others at Save the Children (UK and country offices). The research was designed and managed by Siobhan Warrington who is also lead author of the report; Anna Gormley carried out fieldwork in Bangladesh, Jordan, and the UK and contributed to design and analysis. Hananne Ferdjane carried out fieldwork in Niger.
7. The parents of child contributors are included in this figure.
8. For a full breakdown of research participants by age, gender, type, and country, see the table provided in *The People in the Pictures* full report (Warrington & Crombie, 2017, p. 75).

9. A full list is provided in Appendix 3 of the full report (2017, p. 78).

10. Except for the staff interviews, whereby staff in each country were interviewed with the same set of visual resources (four sets, one per country, produced and used across the 19 staff interviews).

11. Chapter 4 of *The People in the Pictures* presents findings relating to research participants' existing relationships with visual media, and these are further explored in Chapter 7 in relation to how these relationships interact with participants' perspectives on portrayal of themselves and others in Save the Children's content.

12. The research found that the same dynamic of an existing – and unequal – power relationship, between contributor (beneficiary) and image making team (Save the Children) has a significant influence on the contributor's decision to contribute their image and story.

13. Other articles I found, exploring ethical issues and visual methods, focused primarily on what Harper refers to as "auto-photo-elicitation" with photographs taken by research participants themselves, as opposed to the consumption of images taken by others, for example, Ford et al. (2017) and Bugos et al. (2014).

14. Contributors in the UK who participated in the research had, in most cases, seen all of the SCUK content featuring themselves, or their children, before the interview for this research. At the time of the research returning finished content and still images to contributors in the UK, was standard practice among SCUK image making teams.

15. Some may refer to this as assessing and responding to risk.

16. The visual imagery of the film consisted of images of children's journeys to school in several country contexts, with the children all appearing happy and well and optimistic about their school day.

17. All names used within communication of the research are pseudonyms.

18. Available at https://www.noorimages.com/.

19. In 2018, European General Data Protection Regulation (GDPR) made informed consent a legal requirement for those gathering and processing personal data (including images) to be used for fundraising and promotion purposes.

20. GDPR also requires that contributors should be provided with the opportunity to withdraw consent at any point.

21. Bond is the UK network for organisations working in international development, see www.bond.org.uk. Information about the Bond People in the Pictures group can be found at: https://www.bond.org.uk/groups/people-in-the-pictures. Jess Crombie is a lecturer at London College of Communications and an independent consultant; at the time of writing Tamsin Maunder-Betti was Head of Brand and Creative at WaterAid UK.

ACKNOWLEDGEMENTS

I wish to acknowledge SCUK, first for commissioning *The People in the Pictures* research and second for its permission for the author to share her reflections on the research within this chapter. I would also like to thank Dr Edward Ademolu, David Girling, and Anna Gormley for their contributions to this chapter and Dr Rachel Tavernor for providing encouraging feedback on a first draft. My first encounter with photo elicitation was observing anthropologist Dr Julie Flowerday at work in Hunza, Northern Pakistan in 1994. She was conducting interviews about changing landscapes, using photographs taken in the 1930s, and holding the interviews in the same position as the original photographs were taken. I was impressed, and her approach has stayed with me over the years.

REFERENCES

Ademolu, E. (2018). *Rethinking Audiences: Visual Representations of Africa and the Nigerian Diaspora*. Ph.D. thesis. Global Development Institute, School of Environment, Education and Development, University of Manchester, Manchester. Retrieved from https://www.research.manchester.ac.uk/portal/files/77567006/FULL_TEXT.PDF

Ademolu, E., & Warrington, S. (2019). Who gets to talk about NGO images of global poverty. *Journal of Photography and Culture*, 12, 365–376. https://doi.org/10.1080/17514517.2019.1637184

Azoulay, A. (2015). *Civil Imagination: A Political Ontology of Photography*. London: Verso.

Bugos, E., Frasso, R., FitzGerald, E., True, G., Adachi-Mejia, A. M., & Cannuscio, C. (2014). Practical guidance and ethical considerations for studies using photo-elicitation interviews. *Preventing Chronic Disease*, *11*, 140216. http://dx.doi.org/10.5888/pcd11.140216.

Fairey, T. (2015). *Whose Pictures Are These? Re-framing the promise of participatory photography*. Ph.D. thesis, Faculty of Sociology, University of London, Goldsmiths. Retrieved from http://research.gold.ac.uk/22355/

Ford, K., Bray, L., Water, T., Dickinson, A., Arnott, J., & Bernie Carter, B. (2017). Auto-driven photo elicitation interviews in research with children: Ethical and practical considerations. *Comprehensive Child and Adolescent Nursing*, *40*(2), 111–125. https://doi.org/10.1080/24694193.2016.1273977

Girling, D. (2018). *Radi-aid Research: A Study of Visual Communication in Six African Countries*. Norwegian Students' International Assistance Fund (SAIH) Industry Report. Retrieved from https://www.radiaid.com/radiaid-research

Harper, D. (2002). Talking about pictures: A case for photo elicitation. *Visual Studies*, *17*(1), 13–26.

McGee, S. (2005). Report on the review of the code of conduct: Images and messages relating to the Third World. Retrieved from https://www.dochas.ie/sites/default/files/Siobhan-McGee-Final_Code_of_Conduct_Report.pdf

Miskelly, C., & Warrington, S. (2010). Depicting Injustice overseas research: final report. Internal report for Save the Children UK, unpublished.

Sontag, S. (1979 [1977]). *On Photography*. London: Penguin Books.

Warrington, S., & Crombie, J. (2017). *The People in the Pictures: Vial perspectives on Save the Children's image making*. London: Save the Children UK, 2018. Retrieved from https://resourcecentre.savethechildren.net/library/people-pictures-vital-perspectives-save-childrens-image-making

Warrington, S. (2017). *The People in the Pictures: Vital perspectives on Save the Children's image making – Extended executive summary*. London: Save the Children UK. Retrieved from https://resourcecentre.savethechildren.net/library/people-pictures-vital-perspectives-save-childrens-image-making

Warrington, S. (2019). *Putting the people in the pictures first: Ethical guidelines for the collection and use of content (images and stories)*. London: Bond. Retrieved from https://www.bond.org.uk/resources/putting-the-people-in-the-pictures-first.

PART II

POWER AND INEQUALITY

CHAPTER 4

'CAN I TAKE YOUR PHOTO?' THE PRACTICALITIES OF MANAGING INFORMED CONSENT WITH VULNERABLE POPULATIONS

Robert Godden

ABSTRACT

This chapter examines the rights of those who have their photograph taken for use by non-profit organisations (such as non-governmental organisations, charities, and academic institutions) and the obligations of those producing and using these images to not abuse those rights. The chapter focusses on informed consent as the key intersection of these rights and obligations. It examines the specifics of what is meant by 'informed consent' and the importance of the process of it being requested and given. The argument presented suggests, rather than seeing this process as a legal, one-off contract, that it should be seen as a relationship that may evolve over time and, to a degree, attempts to establish a more equitable relationship between those who produce and publish images and those who appear in them. It is suggested that this process can play a role in addressing a number of ethical issues, including the safety of the individual being photographed and how vulnerable populations are represented visually.

Keywords: Informed consent; non-profit; photography; vulnerable populations; participation

Ethics and Integrity in Visual Research Methods:
Advances in Research Ethics and Integrity, Volume 5, 67–91

ISSN: 2398-6018/doi:10.1108/S2398-601820200000005009

Images of vulnerable populations are used by non-profit organisations to expose human rights abuses and injustice (Fig. 1). These images have been integral to research, legal challenges, campaigns, and media work, ostensibly aimed at contributing to positive social change for the populations featured. However, the images produced have also been critiqued on a number of levels. These critiques mainly focus on two issues:

(1) the stereotyping of certain populations and/or portraying them as lacking agency or without dignity; and
(2) the individual's rights in relation to being photographed and how those images are used.

The former issue has arguably attracted the most attention, and continues to do so at the time of writing. The latter issue, although linked to how an individual or population is visually represented, has attracted less debate up until recently. It is this issue that I will focus on in this chapter. Specifically, when working with vulnerable populations, can informed consent policies and practices ensure organisations and photographers respect the rights of those they photograph? If so, what would best practice look like and how can it be practically implemented? With that in mind, rather than just address these questions in theory, it is worth exploring how such a process may be applied in practice by non-profit organisations and the photographers they engage. To do this, an interview was conducted with Smita Sharma, an award-winning independent photojournalist based in Kolkata, India. Her work has been published in various places including

Fig. 1. S., 18, was first trafficked by her aunt when she was 11 and sold her to a placement agency. The agency sold her for domestic work where she was regularly beaten and sexually abused. S. ran away and got help from police and returned home. At 15, S. was trafficked again and sold to another agency. She managed to escape again after six months. Photographed by Smita Sharma.

The New York Times, BBC World, TIME, Channel 4, Elle UK, Wall Street Journal amongst others and screened and exhibited in various countries including the United Nations Headquarters in New York. Smita has worked extensively with survivors of trafficking and sexual violence (Fig. 1).

Throughout this chapter references are made to 'images' and 'vulnerable populations'. Before proceeding it is worth clarifying what I mean by these terms.

The term 'image(s)' is used consistently to refer to a broad category of photographic work that includes: (i) Digital photo files produced within a camera; (ii) photographic negatives produced by a film camera; and (iii) photographs that have been manipulated in post-production software or the darkroom, including the addition of text and/or other artefacts, and compilations of images, such as montage. Where 'photograph' is used, it refers exclusively to the unedited file or negative [(i) and (ii) above].

The definition of what constitutes a 'vulnerable population' varies across international agencies, such as those that are part of the United Nations and the World Bank. For the purposes of this chapter, vulnerable populations are defined by the following characteristics and/or belonging to the following groups:

All children are in some way vulnerable, in that they require the protection and support of adults. As such, they warrant being included as a separate group. However, some are at 'increased risk' and 'in need', and in this case, consideration should be given to the following characteristics: (i) orphaned by the death of one or both parents; (ii) abandoned by parents; (iii) living in extreme poverty; (iv) living with a disability; (v) affected by armed conflicts; (vi) abused by parents or their carers; (vii) malnourished due to extreme poverty; (viii) HIV-positive; and (ix) those belonging to minorities that are marginalised, stigmatised, or discriminated against.

Vulnerable groups of adults include: the elderly; those with mental and physical disabilities; internally displaced people; refugees; those affected by armed conflicts; those living in poverty (including those who are malnourished, homeless, living in inadequate housing, and/or destitute); individuals who are HIV-positive; minorities (based on nationality, religion, language, sexuality, and ethnicity); and, in some societies, women. In regard to the informed consent process, although it may be useful to be aware of the vulnerability of certain groups of people (i.e. susceptibility to harm and types of risk) each case should be judged individually on the specific circumstances of those involved.

The interview with Smita Sharma and analysis presented in this chapter largely focus on images produced by non-profit organisations concerned with human rights and social/economic inequality, however, some of the insights may also be applicable to photographic work conducted by those in other sectors. As such, there is some overlap in the practices explored below, including references to the use of images in the media and research. This may prove useful as freelance photographers often work for both news publications, and non-profit organisations and researchers; and non-profit organisations will often use the media to draw public attention to issues they are concerned with, as well as buying stock photos from media agencies. These relationships are of interest because the differences in ethical standards between the two industries, including when consent is required, and the level of consent necessary for publication. It is also worth noting that the use of images varies considerably within non-profit organisations, whether used for research, legal prosecutions, campaigning/activism, or fundraising. For example, researchers in non-profit organisations often use photographs in their reports to provide visual evidence; whereas fundraisers use images to tell emotive stories in order to elicit monetary donations from the public.

As a response to the economic disruption, the media economy experienced over the last 15 years – characterised by the rapid decline in print sales and the resulting challenge of monetising online news media – more photographers are finding work as freelancers. This places them one step removed from the institutions that would have historically mentored them in the ethics of their work and ensured that ethical guidelines were adhered to. The economic barriers to enter the photographic profession are lower than ever before, meaning competition is higher as more people compete for fewer jobs. For publishers, the speed needed to turn content around has become shorter, meaning time available to check the integrity of a photograph is at times insufficient, especially in regards to consent. In some circumstances, the photographer's word may be the only source of evidence to verify that consent was given. Even when documentation is available, such as standardised forms or videos, it is difficult for those at the editorial end of production to know how the process of gaining consent was conducted. This is important if we are to go beyond consent as a legal issue to one concerned primarily with the participants' rights and protection. Throughout this chapter, I use the term 'participant' instead of more commonly used alternatives, such as 'subject', in order to acknowledge the active role that is necessary for an individual to give informed consent.

Photographers that find work with non-profit organisations can be tasked with photographing vulnerable populations, including children, survivors of sexual exploitation and violence, refugees, and those living in extreme poverty. They may have little knowledge of the issues affecting that population or the risks that those they photograph face. Although more non-profit organisations are now cognisant of the need for gaining informed consent in regards to images, many are playing catch-up in developing and implementing policies. Even where these are in place, some struggle to get compliance across federated structures around the world, where standards and laws differ. Long supply chains, consisting of, for example, the photographer, local partner, domestic office of the international non-profit organisations, and head office of the international non-profit organisations, can make the application and verification of consent polices more complex.

Freelance photographers, having primarily worked for news media outlets, may have a very different idea of what constitutes consent and their obligations to the participants in their photos. News organisations rarely require consent from those featured in their photos, especially those made in public spaces. As mentioned briefly above, another challenge is how to deal with the differing attitudes towards how individuals and populations are pictured, not only across sectors, but also within organisations. For example, fundraisers may take a different view than researchers on what is acceptable due to having diverging objectives and audiences. This can lead to a range of problems, from exposing someone's identity who wished to remain anonymous, making exaggerated or inaccurate claims, or engaging with negative stereotypes and representations of already marginalised populations. Images of children are particularly popular with some non-profit organisations for fundraising purposes, because they are seen as more effective in eliciting a positive response to their public appeals. This can present both ethical and legal challenges, to such an extent that some non-profit organisations are now resorting to the use of child actors in order to address this issue.

What can civil society organisations do to address these challenges? How can they develop and effectively implement informed consent polices and protocols? What can photographers do to ensure they get informed consent in different circumstances, comply with local laws, and not put those they picture at further risk?

IDENTIFYING THE PROBLEM

Since the late 1980s non-profit organisations have been critiqued for the images they use in their work. One of the most widely referenced articles is 'Merchants of Misery' by J. Lissner (1981), which was published in *New Internationalist* magazine. The argument Lissner laid out in the article has become central to the ethical debate around the use of images of vulnerable populations by non-profit organisations.

> The starving child image is seen as unethical, firstly because it comes dangerously close to being pornographic. In my definition, pornography is the exhibition of the human body and soul in all its nakedness, without any respect and piety for the person involved. The public display of an African child with a bloated kwashikorkor-ridden stomach in advertisements is pornographic, because it exposes something in human life that is as delicate and deeply personal as sexuality, that is, suffering. It puts people's bodies, their misery, their grief and their fear on display with all the details and all the indiscretion that a telescopic lens will allow. (Lissner, 1981)

Other writing continued to develop the arguments of Lissner, notably *Imaging Famine* (2005), an online project that presented a range of materials, including essays, critiqued the representation of famine in the Western media and by non-profit organisations; and Alam's (2007) 'The Visual Representation of Developing Countries by Developmental Agencies and the Western Media' that highlighted the dominant role of the Western photographers in visualising the world and how this lack of diversity distorted the understanding of issues such as poverty amongst Western audiences. More recently, articles like Pixley's (2017) 'The Western Gaze: On Photojournalism and Challenging Harmful Representations' and Mishra's (2018) 'The Great Upheaval: Can the digital revolution potentially shift the power dynamic in photography?' have expanded the focus from how certain populations are depicted in images to look at the underlying causes of this, including an exploration of the demographics of those with the power to produce and disseminate images.

These critiques largely focus on the negative representation of vulnerable populations, through images that depict people as lacking agency and/or without dignity. More recently, articles such as those by Mishra (2018) and Pixley (2017) have characterised this form of racial stereotyping as the 'colonial' or 'western' gaze that has its roots in European colonialism and the role photography played in cataloguing, dehumanising, and controlling populations. The othering of these populations through photography was one of the tools used by governments to justify their imperialist policies. The critique claims that such a gaze continues today in Western photography (in particular, through photojournalism and non-profit organisations), in part due to the lack of diversity in the profession, including at the managerial level. The result is a distorted visual representation of certain parts of the world, the people that live there, and the events that affect their lives.

This 'colonial gaze' critique resulted in a self-examination by National Geographic magazine in their special edition devoted to the issue of race. In an article titled, 'For Decades, Our Coverage Was Racist. To Rise Above Our Past, We Must Acknowledge It', the Editor in Chief, Susan Goldberg (2018), wrote about how the magazine's coverage, including images, had been historically racist. As part of this self-examination, the magazine commissioned John Edwin Mason, a University of Virginia professor specialising in the history of photography and Africa, to examine the images in their archives.

In addition, non-profit organisations have faced similar criticism that their use of dehumanising photos in fundraising appeals for humanitarian relief has resulted in public apathy – so-called 'compassion fatigue' – amongst Western audiences. Compassion fatigue can be defined as the numbing of the public's capacity to emotionally respond to humanitarian crises due to the repetition of images (and news articles) depicting such events. This phenomenon is largely attributed to the populations of Western Europe and North America, and is usually linked to emergency relief appeals by non-profit organisations. In an article for *The Guardian*, Gabbert (2018) quotes psychologist Charles Figley as defining compassion fatigue as, 'a state of exhaustion and dysfunction, biologically, physiologically and emotionally, as a result of prolonged exposure to compassion stress'. Others, such as World Press Photo Foundation's Director of Programs and Outreach David Campbell, have challenged the compassion fatigue argument on a number of occasions, most notably in 'The Myth of Compassion Fatigue' (Campbell, 2014). Campbell's challenge largely focusses on the lack of evidence to support the phenomenon, including academic studies on particular humanitarian crises and statistics for the amount of money donated to these humanitarian appeals.

The debate surrounding compassion fatigue continues to generate sporadic arguments for and against, but rarely becomes a hot issue of debate. However, the critique of how non-profit organisations and media organisations visually represent certain populations in a negative and clichéd way continues to attract attention and gain support. This has, to a certain extent, resulted in a change of policy amongst a number of non-profit organisations. One example is the 'Code of Conduct on Images and Messages' (CONCORD, 2006). Although a broad code, it outlines commitments to the selection of images that 'respect dignity', avoid stereotypes, and have the 'full understanding, participation and permission (of subjects' parents/guardian) of subjects' – effectively establishing a commitment to informed consent in relation to the production of images.

Informed consent is linked to, and in some respects, flows from, the issue of negative visual representation, including the use of stereotypes, tropes, and demeaning images. Central to this is the unequal application of ethical standards in how individuals from certain places and backgrounds are visually represented in comparison to others – that these standards are applied unequally depending on the race, ethnicity, sexuality, nationality, gender, or class of the individual. The argument goes that not only do we need to treat people equally, but also they have rights in regard to their visual representation and how their image is used. As such, their participation in being photographed should be conditional on their consent.

This changing attitude towards the rights of those who participate in the images produced and used by non-profit organisations has resulted in the popularisation of informed consent as a practice. Though largely positive, it raises a number of questions. First, what do we mean by 'informed consent' in relation to the production and publication of images? Can we agree on a set of criteria in order to establish a standard definition (and would such a definition be desirable)? Second, what degree of information is required by both sides in order to: (i) satisfy 'informed'; and (ii) allow us to be secure in giving/receiving 'consent?' This inevitably leads us to the issue of risk and our ability – both as photographer/publisher and participant – to realistically assess this. Third, one criterion that is often included in definitions of informed consent is the ability of the participant to withdraw consent. This raises the challenge of how, in the digital age, we can control images once they are online, and so thus honour this requirement.

Children: A Special Case for Protection

Attitudes and laws related to child protection have evolved internationally over the last 20 years, largely driven by the entry into force of the UN Convention on the Rights of the Child (1990), the most ratified human rights treaty in history. Although each country has responded differently, there is widespread recognition that laws to protect children from exploitation and abuse are essential. Governments have responded to a mixture of pressure from progressive politicians, civil society organisations, and changes in public attitudes. Industries and professions have adapted and adopted mechanisms to ensure they comply with changes in the law as well as changes in attitudes.

However, questions continue to be raised on how children are visually represented by non-profit organisations. One recent example was a critique by Benjamin Chesterton, Production Director at Duckrabbit (2018), a UK based video production company, on Twitter questioning the use of images of children by the humanitarian non-profit organisation, ActionAid, in a fundraising advertisement used on social media. The advertisement purported to tell the stories of children who had been sexually abused, including rape. It was revealed at the end of the piece that child actors had been used. Chesterton questioned the consent process for the child actors' parents and suggested the advertisements were misleading to the public. In order to comply with their own ethical policies on the use of images – not to publish photos of children who had been abused – ActionAid's solution created its own ethical questions: (1) Was it disingenuous to not state clearly enough that child actors were being used? By not doing so it gave the appearance that ActionAid wanted people to believe the children featured had really experienced abuse in order to move them to engage with the campaign, yet if the audience believed this then would it in some way contradict ActionAid's own policy? (2) Should child actors be represented as being victims of child abuse? Could this also have a negative impact on the child, and what procedures had been in place to get consent from the child actors and their parents/guardians?

These questions not only raise issues of representation, as outlined above, but also whether the children are being re-exploited and/or put at risk through the production and dissemination of these images. This can be linked to the added layers of complexity in getting consent where minors are involved.

DEFINING 'INFORMED CONSENT'

Photographers and non-profit organisations are facing increased scrutiny in regards to their ethical conduct in both how their images are produced and the type of images they are producing. A critique of such an image will often elicit a response that those within the frame 'consented'. This would appear to indicate, at the very least, that consent is perceived as a way of legitimising the ethical credentials of an image in regards to the cooperation of those pictured (though this still leaves the originators of the image open to criticism on other issues, such as representation). Whether or not a consent process occurred at all, was reasonably 'informed', or was part of a systematic policy, can only be assessed based on the evidence presented. As such, if we are to understand and assess how individual photographers and non-profit organisations approach informed consent we need to know what specifically they mean by it and what process they follow to gain it? Transparency in this regard would go a long way to progressing understanding of the issue and adoption across more organisations. For example, non-profit organisations could publish such policies and the related materials.

Although fairly standardised across academic institutions in relation to conducting research, definitions of what constitutes informed consent in the production and use of images will vary between non-profit organisations. The principles outlined below are loosely based on those developed by Amnesty International's International Secretariat for research purposes, and share similarities with those used by other institutions. They are included as a basis to understand the practicalities of implementing an informed consent policy across an organisation and by photographers working in the field.

The principles of informed consent are that it must be: (i) understood; (ii) freely given; (iii) informed; and (iv) specific. It is important that effort is made to ensure all the above factors are present when seeking consent. Consent is a pro-active process and should never be assumed due to other forms of previous cooperation. It is also not a one-off event but an on-going process, and as such should be periodically refreshed (as the circumstances of individuals may change over time). Consent to be photographed at one time does not mean consent to be photographed in the future, nor does it mean that consent cannot be withdrawn, or that consent applies to all situations and locations (e.g. agreeing to be photographed at home does not mean agreement to be photographed at work).

The principle of informed consent implies the participant should receive explanations in simple, jargon-free language, as to the following:

- The identity of the information collector (i.e. photographer or researcher), along with a brief explanation of the commissioning organisation. Examples of work by the photographer or commissioning organisation can be shown to potential participants. For example, small cards can be made in relevant languages for easy explanation.
- Why the photographs are being taken and their intended use. This should include not only an explanation on how the photographs will be used but also the type of materials and digital platforms they could appear on. The participant should have the opportunity to object to particular uses and/or forms of publication, particularly where these could put them at risk.
- Details of the potential risks to participation in the process. This should be a two-way process in recognition that the participant may be more aware of the risks their participation exposes them to than the photographer and/or organisation.
- The meaning of confidentiality and how it applies with special emphasis on the fact that the person photographed can request his/her identity to be kept confidential. When anonymity is requested, special attention must be paid to a wide range of possible identifying factors beyond the participants face. For example: items of clothing and/or jewellery, locations (interiors and exteriors), scars, birthmarks, etc.
- Contact information so that the participant can reach the information collector (and vice versa).
- Details on how long the information will be used, and how and where it will be kept or stored.
- Reminders that the participant can cease participating at any time (i.e. withdraw consent), and request that his or her information and images/video be destroyed, whenever feasible. The ability of an organisation to remove an image from the internet once published is extremely limited. Although it can remove versions from its own files and websites, and stop using it in their materials, images already published may be replicated across multiple sites which the commissioning organisation has no control over. In addition, it is worth noting that once images are published there is often little to stop others using or altering them in order to give a different meaning than intended.

Consent signifies the approval by the participant for the photographs, video, and/or audio to be used as explained. Consent can be given with or without limitations. It must therefore be specified whether the whole series of photos can be used, including the identity of the participant, or whether the information may be used on condition that the identity of the participants is kept confidential. The participant may deem some parts of their story/photos confidential, and others not; this should also be clarified and recorded. Care must be taken in the drafting of captions/accompanying text to ensure they are accurate and include only information that was agreed to as part of the consent process.

Children

When photographing children, an additional level of consent is required in order to protect them. Consent should form the basis of the production and dissemination of images of children that respects and protects their rights to express views, to information, to privacy and to protection from violence and abuse. Photographers should, wherever practically possible, seek consent from the child as well as their legal guardian when carrying out their work.

Consent can only ever be considered to have been freely given if the child believes they are able to decline, giving them a meaningful choice. This choice may not be as immediately obvious to children as it is to adults, thus, must be explicitly communicated. Photographers should be aware that children may believe there could be benefits to them participating, thus, it is important that expectations are clearly managed. How an individual or community benefits from being photographed can never be precisely quantified. As such, photographers should only state how they intend to use the photographs rather than perceived or possible outcomes. It is worth noting at this point that this principle also applies to seeking consent from adults.

With children, photographers need to determine they are able to understand why they are being photographed and the potential consequences of participating and their photos being published. In obtaining consent from children, it is important to understand that a child's cognitive and communication skills are different from adults. It is important to assess these skills, including their maturity and ability to make decisions, when determining whether informed consent can genuinely be given. It should be noted that age alone should not be used to determine children's ability to give consent. Other factors that impact maturity should be considered and decisions made on a case-by-case basis. In some jurisdictions, children below a certain age may not be legally able to give consent. Photographers should check with laws in the country they are working in, the country they are usually based, and the country where the photos may be published.

It is important to highlight that when seeking consent from a child it is not enough to get the go ahead from their guardian (nor just from the child if no guardian is available). The rights of the child must always be respected. In addition, it is important not to assume that the child's guardians always have their best interests at heart.

Risk

Consent can only be considered informed if the risks of cooperation are explained and considered. A risk assessment by the photographer and/or the commissioning organisation, taking advice from relevant experts, is essential in being able to help participants make informed decisions on whether to cooperate or not. The assessment should take into account not only the risks to the participants from publication (i.e. publicly telling their story), but also the act of photographing itself (e.g. being seen to be photographed), as well as the association with the commissioning organisation. Factors to consider include, but are not limited to: (i) Personal/family – what could be the impact on the subjects relationship with friends and family seeing the photos or being associated with them?; (ii) Social/cultural – what could be the impact on the participant (and those close to them) where they live and in their local community?; and (iii) Political – what could be the impact on the participant in regards to reactions from the state (e.g. police and government)? In answering these questions, it is also important to assess the probability of the negative outcome occurring and what level of risk is acceptable.

Duty of Care

Informed consent includes an understanding of any risks involved in being photographed and having the photographs published. Even if the photographer decides that the participant understands the risks involved and still agree to consent, the photographer must not transfer responsibility for the risk to the participant. The participant's vulnerability to risk will be influenced by their situation. In the case of children who are labourers, orphans, homeless, or in institutions/detention risk may be heightened. In addition, gender, armed conflict, social violence and poverty can increase the risk for all individuals and their ability to access protection mechanisms. When assessing risk, the photographer should consult with relevant experts, and if they conclude the risks are too high, the photographer should not proceed with photographing the individual(s) even if the participant has consented (i.e. the risk tolerance of a participant is higher than that of the photographer and/or commissioning organisation). With an adult this becomes a judgement call that the producers of the photograph must deal with – whether to respect the wishes of the individual to have their image published (freedom of expression) or to decline in order to protect them and/or the organisation (do no harm). There is no easy answer to such questions, which would have to be dealt with on a case by case basis. Declining presents the added risk of being criticised for being paternalistic and/or censoring an individual with their own agency to decide for themselves what level of risk they are willing to take. Whereas, accepting a higher level of risk opens up the photographer and/or the commissioning organisation to the criticism of failure to protect if the participant faces negative consequences linked to their photo being taken or published.

IDENTIFYING THE SOLUTIONS

Any organisation that wishes to ensure the images it commissions and uses are gained via informed consent will need a policy that is robust and detailed in its ethical standards, as well as having practical guidelines for those who make the photographs.

Practical Experiences from the Field

When considering how a photographer can get consent in an informed and ethical way, we need to ensure that the process is simple and practical. In order to gain insight into how these policies might be developed and work in practice I spoke with Smita Sharma, a freelance photographer based in Kolkata, India.

The conversation with Sharma was conducted via Skype on 19 February 2019 and was based on a number of communications with her over the last four years. Sharma's work has focussed on gender based violence, trafficking, and sexual exploitation. This focus has necessitated the development of formal and informal methods of securing informed consent. These methods go beyond the form-signing, legalistic action of asking if she can take someone's photograph. What Sharma revealed was that informed consent requires a commitment to respecting the wishes of the participant, and a pro-active attitude in adapting to each individual scenario.

Sharma is an award-winning independent photojournalist. Her work has been published in various places including *The New York Times*, *BBC World*, *TIME*, *Channel 4*, *Elle UK*, *Wall Street Journal* amongst others and screened and exhibited in various countries including the United Nations Headquarters in New York. She is a Getty Images grantee, an International Women's Media Foundation (IWMF) Fellow and recipient of The Washington Post Award at the Eddie Adams Workshop. She received the Indian of the Year award in 2017 for her contribution in the field of journalism, Exceptional Women of Excellence award by Women Economic Forum in 2018 and won the One World Media awards for her film 'Rebels with a cause' (2019). She graduated in photojournalism and documentary photography from the International Center of Photography, New York and has a postgraduate diploma in Journalism from University of Pune, India.

The following conversation has been edited for clarity and length.

Robert: I would like to focus on the practicalities of a photographer getting informed consent from an individual in the field. I want to learn more about your personal experience, with the aim of understanding how we might produce guidelines or training for photographers committed to integrating this into their practice.

Smita: As a photographer who deals with very sensitive issues, such as sexual violence, child marriage, or trafficking, I have always been acutely aware of my responsibility to those I photograph. That includes not only the survivors but also the families who have been effected by these tragic events. When you meet these people, your first priority should be to inform them why you are there. Be very honest with them regarding your intentions, your purpose, etc. This includes explaining how their story is going to be used.

Even people in South Asian villages have mobile phones and the internet, so if you are working on something that is sensitive – and the people do not want to be seen – you have to respect that. For example, I was in Bhopal earlier this year working on an assignment for a non-profit organisation. The issue being covered was domestic violence. I met the girl who I was to photograph, and I asked her if it was OK. She understood what was being asked and specifically told me, 'I don't mind you taking my photograph but I don't want anything coming out on Facebook'.

Many people do not care what you do with their photo, but some are very specific in regards to where they do not want their photo to be published. It is a very individual, personal choice, and so we must adapt ourselves each time we ask for consent.

Robert: I think that is a very important point. Sometime I feel that consent is reduced to a form filling exercise, where the objective is to get a signature.

I want to go back to what you said about explaining your purpose. Before you even ask someone if you can take their photo and get consent, you need to explain who you are and what you are doing. This is fundamental to the whole process of informed consent. This information allows the individual you wish to photograph to come to a decision. Do you find explaining your purpose easy?

Smita: I would say it differs between people and situations. However, no matter what a person's level of formal education they are not stupid. I think that if you explain that the reason for you being there is to document a story that may not directly help them but could more broadly expose the issue, people do not mind sharing their stories. Many times people will say to me, 'What has happened to me has happened, but I hope it doesn't happen to others'. It can also be empowering for them, they feel that their story is important, and provides solidarity with others who have experienced similar things.

I was working on a very sensitive story in India. This girl was trafficked and disappeared for a few months. The parents had been distraught and filed a report with the police. She was eventually rescued in a police raid. In small villages in India people talk and so the parents made up a story that their daughter had gone to stay with a relative. When she came back they continued with the pretence. If people found out she had been trafficked it could do great harm to her reputation and future prospects. When I went to photograph the family in the village, the mother was very concerned and did not want people to see me with a lot of photographic equipment and ask why I was there. So, I took very little equipment, just one [camera] body and one lens in an ordinary bag so as not to look like a journalist or photographer. When I was there the neighbours became curious about my presence and asked me questions. To protect the family's story I went to the neighbours and over the next few days talked to a number of girls about menstruation, explaining that I was there to do research on this issue. As soon as I started talking about such things, the men became embarrassed and left. I also took the portraits of the girls I spoke to. This successfully deflected attention from the family and protected their story.

Robert: That is an interesting strategy as it speaks to the risk posed by the presence of the photographer. I am also reminded that there are many factors that an individual has to weigh-up when deciding on whether to give consent or not. Rarely is it a risk-free decision. Should we consider giving consent as part of an agreement, the other side of which includes protection?

Going back to the issue of explaining purpose, as a freelancer do you bring any materials to illustrate the publications you are working for or how the images might be used?

Smita: Yes, I have these uploaded to my phone. I have a folder with examples of my work, to give them an idea. I also show images I take of them on the back of my camera so they can see how they are being visualised.

Robert: How does your role, as a photographer, influence the power dynamic between you and those who participate in your photos? For example, if you are working for an international media outlet such as *the New York Times* or a large non-profit organisation, how does this impact how they view you? Does being a middle class professional working in a poor rural area influence how people view you? How do you deal with the real possibility that your status can influence people's decision on whether to consent or not? Do you think they may feel pressured or obliged to cooperate, where ordinarily they may not have? Equally, this could be the case if you are accompanied by a staff member from a non-profit organisation which runs programmes in the area which local people have benefited from.

Smita: I think it is important to realise that we are nothing without the people we photograph. It may seem obvious but I think sometimes people forget this simple truth. They are doing us a favour and we always need to remember that. I have seen some journalist and photographers that think because they have travelled a long way, and they need to get a story, they can be pushy or arrogant. It is important to be humble, and to not use any inherent status you may have to influence their choice. But it is hard, often we are working with people who have been marginalised or discriminated against, social hierarchy is difficult to neutralise.

Robert: It strikes me that a key element [to the informed consent process] is the attitude you bring as a photographer to the situation as this shapes the whole process. It is not just a tick-box process, but that you genuinely believe that those you photograph have a right to understand what you are doing and what will happen to the photographs you make of them.

If this is not the case, then everything that comes after will likely fall short of informed consent. Although we can have the practical tools and forms, it is fundamentally a commitment, by both the photographer and the organisation they are working for, that they genuinely committed to the process.

I would like to ask you specifically about getting consent to photograph children, particularly the process of approaching their guardian, but also how the opinion of the child can be included, based on their age. There are a whole range of developmental stages, so a six year old has a very different level of understanding than a fourteen year old. What is your process?

Smita: The first thing to consider is their age, as you say, as this affects their understanding. You do not need to explain everything to the child but you do to the guardian. In India, there are specific laws you need to abide by and anyone who wants to photograph children here needs to understand these. The first is Section 73 of the Juvenile Justice Act, and the second is Section 23 of the Prevention of Children from Sexual Offences Act. Particularly Section 23(2)[1] is very important. In addition, I would also include the 'UNICEF Guidelines for journalists reporting on children' as that is universally useful. As a professional photographer, you need to do your research before going out in the field, and I feel many photographers do not do this in enough depth. I continue to see very sensitive stories on rape and sex trafficking showing the faces of the survivors and their children. This can be very dangerous.

Robert: How do you manage the issue of expectations, of perceived benefits by the participants of taking part in the photographic process?

Smita: I have been in situations like this. Often people feel like they are not listened to and because I am there asking them questions, they are grateful that someone wants to hear their story. Sometimes it is therapeutic for them, and that can be enough. But there can also be other expectations, financial or legal. I always explain what my role is, to inform, through words and pictures, and I do not have power beyond this. It may help bring about change, but I never oversell this. I can facilitate others to help them but I will not do that directly as a journalist. I cannot pay them in any way for their help. It will look like I am paying them for information, and I explain that clearly. People understand.

Robert: I believe we have an obligation to understand what the risks of consenting are to that individual. What they perceive are the risks to themselves and their families, and the wider context. I want to divide this into two areas for consideration:

(1) The risk created by you being there taking photo, for example, the impact of your presence; and (2) the risk created by the publication of the images.

What is your experience of these two elements of risk?

Smita: In regards to publication, if I am working on a project that I do not know where it will eventually be published, I inform people that it could appear both nationally and internationally. Sometimes people will then tell me they do not want it to appear in a local newspaper, which I can then take into account. In this respect, social media is very important. You have to be very careful. You have to think about how the images can be used beyond your original intent. That is why I take hiding people's identity very seriously. Although some requests remain within my control and can honour them, I cannot do much to prevent others using the image in ways that breach that request.

In regard to the risk associated by my presence, village headmen [in South Asia] can be a big problem. They are like a king in their village and do not want negative news coming out. People have mobile phones and use social media, so the moment you as an outside appear the news circulates very fast. I had an experience when a motorcycle gang came and started asking me why I was there and what I was doing. The villagers came and defended me. I remained calm and tried to defuse the situation. I am always very open and answer questions. I always try to take the attention away from one particular family so they do not get in trouble. Knowledge of the social and cultural ecosystem is incredibly important in navigating these types of situations.

Robert: Much of your work has been with individuals who experienced traumatic events, whether being trafficked or surviving sexual violence. One important consideration when working with such individuals is what will be the emotional impact on them of re-telling their story? If the process is not handled correctly it could be re-traumatising, though equally, if done correctly, it could be therapeutic. How do you manage that process so that it has a positive, or at the very least, a neutral impact?

Smita: I think it is about patience and time. You cannot rush through things to get the whole story in a few hours. That is not fair to someone who has gone through such a traumatic experience. You cannot just immediately ask them, 'What happened to you?' So, spending time with them, getting to know them, who they are, sharing your own experiences, make it more of a two way process. Only once I have built up this relationship do I ask, 'Can I ask you some questions? Are you OK with that? Do you mind telling me what happened to you?' So far this has worked, but I never assume that they will share these events with me.

Robert: I am interested in what you said in regards to time. It strikes me, in the same way as 'intent', that it is an unseen, yet key, component of informed consent. It is about taking time and not forcing the individual to proceed in a way they may be reluctant to. For example, if a photographer has only a short space of time to get images, they themselves may feel obliged to force the issue of consent where it may otherwise not be forthcoming. They may not have the time or patience to develop that relationship that you just outlined, so that when we talk about informed consent we might also be talking about the time it takes in order to really get that. We also need to consider, in regards to the issue of trauma, that if consent is rushed or forced, although it is given it may end up damaging that individual.

Smita: Yes, that is a significant issue. If you try to cover an issue in just two days, especially something that is really sensitive, I think it is not feasible. If you want six stories, testimonies, in two days, it is not fair on photographer or those they photograph. It will fundamentally impact the consent process.

Robert: I want to ask you about this process you have laid out – explaining your purpose, mitigating risk, ensuring you do not retraumatise anyone. You obviously have quite a bit of experience doing this, but at some point this must have all been new to you. How did you go about learning how best to proceed, what was your methodology when first faced with getting informed consent?

Smita: I studied at the International Center of Photography but they do not teach you about informed consent there. You learn it in the field. I was extremely nervous when I started out, working on these sensitive issues. I gave a lot of time to each of the stories I worked on. I did a lot of research before going to the field. When I was out in the field I never went straight to asking people to tell me their story. I would let things evolve slowly. It took time. I was rejected a couple of times, but I also had some success, so I learnt slowly.

Robert: So this was an experiential learning process rather than being taught in the classroom?

Smita: Yes, absolutely.

Robert: Although that appears to have worked for you, I am conscious that this might not be the best way for everyone and more support from organisations and educational institutions could be very helpful in this regard.

I want to get into the practicalities of recording informed consent and how you go about this. I know you may have different methods depending on the situation. How might you do it for a personal project, for a publisher, or a non-profit organisation, where these organisations may have their own processes for getting and recording consent? What are the pros and cons of written consent, audio consent, forms, etc.?

Smita: If I am working for a non-profit organisation, it is usually with a researcher or someone on the ground who works with me, so the consent is usually taken care of by them. They explain everything, and then I do the photography later on. Regarding publications, they usually have their own model releases [consent forms], some do and some do not. Some say getting consent is up to me and do not provide any institutional materials. Some have detailed forms. I generally carry my own model release forms, sometimes in a local language. I give them a copy if they want.

Robert: What is your experience of asking people to fill out and sign forms? Are there ever trust issues? Have you tried using video or audio to record consent, and if so what is your view on this in comparison to written?

Smita: I always use written consent, though in addition I might do video consent if it is a very sensitive or complex topic. This then allows me to ask more specific questions.

Robert: Each individual and organisation has their own consent forms and processes, and these vary from being quite vague to quite detailed. Such diversity in approaches would suggest it is not enough to claim 'informed consent' was given without further detail on the specifics of the process.

You also mentioned people saying they do not want their image on Facebook or local newspapers. Do you find that organisations you work for have that level of detail or allow you to record that level of detail in their consent procedures?

Smita: Generally, they are detailed but do not always cover things like local newspapers. Most would only distinguish between national and international. If there is an individual request, like not putting their image on Facebook, I make sure it is recorded.

I was working for a non-profit organisation on a campaign in informal settlements. The campaign was launched on social media and on their website there was an image of a girl who was four years old. The text said something about 'your home not being safe as you are abused there'. The language was bold and seemed to insinuate that the girl was abused, but she was not, she was just living in a slum. But a small '*' took you to the next page and explained the girl was a child actor. Some saw this image but not the '*' and so were a bit shocked that I would allow the non-profit organisation to show the identity of this girl. They wrote to me privately. I was shocked. It was never explained to me that it would be used in this manner. I do not think their intentions were bad; they just did it in a bad way. I contacted them and told them what I thought. The first response was very generic, telling me they take these issues seriously and have policies, etc. That annoyed me as I took the photos! I wrote back and told them that I had been involved and their usage of the photo is unethical and impacts my reputation. That this use of an '*' was not good enough as most people will not even see it, and those that do will not click on it. They apologised and changed the page.

Robert: We have discussed about protecting an individual's identity and that as photography is a visual medium it can be quite hard to take photos of people who do not want to be recognised. That is an issue that photographers have to deal with creatively. But I wonder if you can say something about protecting anonymity, what are the implications for those who are photographed?

Smita: I want to talk about shame. In certain communities and countries, shame is a big thing that is not always understood by many Western photographers. What it can lead to. I say this as my cousin was bullied in school after reporting sexual harassment, which the school tried to suppress in order to protect its reputation. The teachers then targeted her, trying to blame her for the harassment. She felt abandoned and humiliated, and this went on for a number of years, eventually culminating in her committing suicide.

It is a big deal. Things can get complicated. There are many layers to what can happen. We have a responsibility, and if anything happens to them as a consequence of that photo we have to be held responsible.

There was a German photographer who did a project on brothels in Bangladesh. She received some awards and recognition, the work was seen widely. She wrote to me privately and said that some people believe she exploited the women she photographed. But she said she went there with a local non-profit organisation and they said it was OK, plus the women did not mind either. I told her that you come from a certain country and background and you do not understand the culture of this other place. The photo you took of the young woman with the older man, what happens if tomorrow she is rescued and goes home to her village? Do you know what consequences this might have for her? Her community knowing she worked as a sex worker? No one would want to marry her, and no one would want to talk to her. Her entire family could be ostracised. They might not be invited to weddings and other community events. I have seen this happen, how a family is shunned by an entire village because their daughter was raped. You need to know the cultural differences before you go and photograph. You need to do your research.

Also, on the issue of anonymity, I did a project with Human Rights Watch where the researcher was worried the individuals I was to photograph might be identifiable by their clothes, even with their faces obscured or in shadow. So we went to the market and bought lots of colourful cloth to wrap them in, which turned out to be a creative solution.

Robert: It strikes me that for those coming from outside the geographic and cultural context of where the photographs are to be made there is a steep learning curve, and the cultural distance makes it more difficult for them to understand the nuances of a situation. Their responsibility to try and understand the situation is thus heightened. This also applies when working with a local non-profit organisation. There is sometimes a lack of appreciation that you cannot just rely on a non-profit organisation saying it is OK to take someone's photograph. Not all non-profit organisations are equally professional, and not all understand the concept of informed consent.

Smita: Yes, some of the local non-profit organisations, when they send a photographer out with their people, they are usually field workers. You cannot assume they appreciate all the risks and requirements of informed consent.

Robert: That also speaks to the issue of representation and dignity. Let us use the example of photographing sex workers. That there is this othering or exoticising that tends to ignore the consequences to those individuals. To photograph them in a way that others would not be photographed or even how sex workers in Western countries would be photographed. This moves us slightly away from consent but I feel it speaks to the wider responsibility that photographers have. That when you get consent from someone, there is an implicit assumption that they will not be presented in a derogatory, undignified or misleading way. Although this is never explicitly said, either on the consent forms or in conversations, I do not think people would agree to be photographed in a way that they found degrading.

Smita: I think it is not necessary to show people using pity to provide insight into a problem. I think the best thing you can do is to show them with dignity. Even if they are in the worst situation, everyone has a right to be represented with dignity. We do not have the right to take that away from them.

Robert: There is a lot of talk about presenting people with dignity, but to play devil's advocate, there are situations, such as documenting human rights abuses, when people are not in dignified situations or abuse is on-going. Some would argue you need to portray this to shock people or communicate how grave the situation is.

Smita: I think it is about balance. What is happening to children in Yemen, showing a child who is malnourished, I think it is important to show that. Whereas if you are showing someone who was with ISIS, has now escaped, and is pregnant, then why is it necessary to show her naked? I do not think that is necessary. It is important to show the real situation but not strip someone of their dignity. It can be hard to articulate.

Robert: Yes, it is something we feel we instinctively know but find it difficult to define. I wonder how this issue might be addressed? How we can deal with the ethics of representation. How might that be addressed? You mentioned earlier that you show people their photo on the back of your camera so they have an idea of how you are photographing them. This reminds me of a process we used to produce photographs for a book project a few years ago. We gave the participants an opportunity to see the photos we wished to use in the book and that they could flag any they objected to. We would then have a conversation with them about that image and whether we should remove it.

Obviously, this process was in part determined by the circumstances of how the book was produced. We had time to do this and the ability to contact them months later. However, I think about how an individual who agrees to participate in being photographed can have some power over not only consenting, but also the selection of the photos. I know there are many photographers and editors who would not like this, however do you think there is something to be said about having more equal power relations between those make and distribute photos and those who are photographed?

Smita: No, I would not do this. I would not give the power over visual content to the people I am photographing. Telling me what they like and do not like. They do not understand visual communications or what makes a powerful photograph. So, if I have explained to them what I am doing and they understand, then they need to respect and trust that I am not going to show them in a bad light. I will show them examples, and for those who wish to remain anonymous I will show them how I am obscuring their identity. In the end, I am not going to be the one who chooses it, it will be my editor.

Robert: OK. That takes me on to another issue about how the photos would be used. For example, specific requests for which platforms the images can and cannot be used on. How much power does the photographer or publisher have to control the image once it has been published? Once the image is on the internet, although you may honour your commitment not to put it on a certain platform, there is very little to stop others doing so. Do you feel that can be explained adequately and understood? Also, withdrawing consent – it is usually presented as a one off act that you sign a form and that is it. What is the power of an individual to change their mind, and the power of the publisher to do anything about it? What is our power to control the image and thus honour consent?

Smita: Well, as a first step, I am very careful in selecting the photos I send to my editor or the non-profit organisation. I am particularly mindful of images that may present risk if they get circulated in the wrong way. Or if I do have to submit it I would write my concerns about the images that I think present some risk. I have taken out many photos from my edit, even with consent, due to not feeling comfortable with them. There is nothing worse than feeling you endangered someone through a photo you took.

There was one photo of mine that *TIME* magazine wanted to use as the main photo for an article, but I had my doubts and once I had talked to them they decided not to use it.

How do you control social media? What if someone takes a screen shot and posts it on Facebook or anywhere else the subject said not to? I do not know how much you can control that?

Robert: I wonder, depending on the circumstances, whether there might be some power with the photographer or publisher to retract an image, remove it from their sites, not sell it or publish it anymore, remove it from their website etc. Do you find ways to keep in contact with the people you photograph so they can be in contact with you if necessary?

Smita: Yes. I take their phone number. Or if they do not have a phone then someone else who has one in the village or a contact at a local non-profit organisation. I always try to keep the gates open.

CONCLUSION

In 2018, Rights Exposure partnered with the Journalism and Media Studies Centre at the University of Hong Kong and the Explore Program at the World Press Photo Foundation to organise a one-day conference entitled 'The Ethical Image: Challenges in Visualising a Changing World'. The conference brought together experts from academia, media, and non-profit organisations, to discuss the ethical challenges in making and using images.[2] One outcome of the conference was a recommendation from the panellists to work towards a practical guide on informed consent that could be adopted by non-profit organisations and photographers.

The development and implementation of a policy such as the one described below presents serious challenges to organisations. Photographers working to such policies would be required to adapt their practice in order to comply with the consent protocols of the commissioning organisation. However, it can be argued that not only would such a situation protect the rights of the participants, but also safeguard the photographer (and commissioning organisation) from legal challenges and accusations of unethical behaviour. As such, it could be reasonably argued that all those who use photographs of vulnerable populations in their work would benefit from not only understanding the concept of informed consent but applying a policy similar to that outlined in the framework below.

Proposed Framework of Actions to Inform Ethical Policies on Image Gathering

- Establish an organisational definition of 'informed consent'. Provide detail on how consent must be (i) understood; (ii) freely given; (iii) informed; and (iv) specific.
  - Ensure the definition provides for consent to be withdrawn. This should include instructions on the practicalities of both the organisation's staff and the participant being able to contact each other, including periodic check-ins to 'refresh' the consent (e.g. every two years).

- Establish an informed consent policy that includes:
  - The need for a risk assessment. Provide details on what areas such an assessment would likely cover (e.g. physical harm, reputational harm, economic harm, legal consequences, etc.).
  - The organisation's commitment to the 'Duty of Care' principle to ensure participants' risk tolerance does not unduly endanger them or the organisation, whilst respecting their right to freedom of expression.
- Develop practical materials that allow staff and photographers to implement the informed consent policy. These should include:
  - Standardised text that explains in simple, jargon-free language what the organisation does.
  - Guidelines on recording consent, including instructions that consent must be presented in a language of which the participants have a good command.
  - Standardise forms for photographers and staff to use for recording consent in writing. It is however worth noting that the use of written consent forms may not always be the most appropriate method of recording consent. Such forms may put pressure on the child or guardian by creating a false sense of obligation, as well as giving the impression that once given consent cannot be withdrawn. In some places asking people to sign forms will be met with suspicion, as well as presenting challenges where literacy is limited (especially with young children). Recording consent using video on a mobile phone may provide a more verifiable and accessible option. However, in some jurisdictions there is a legal requirement to get written consent.
  - Instructions on how video consent must be recorded and a list of questions to be responded to by the participant.
  - Standardised list of consent options, including; (i) circumstances photographs can be taken (e.g. location, timing, activities, etc.); (ii) distribution (e.g. hardcopy publications, social media, news media); (iii) anonymity; and (iv) how the image can be used (e.g. for research, legal evidence, fundraising, media work, campaign activities, etc.).
  - Standardised check list for photographers reminding them to explain the following to participants in their photos:
    - Who they are?
    - Why they are there?
    - Why the photographs are being taken and their intended use?
    - Details of the potential risks to participation in the process.
    - The meaning of confidentiality and how it applies with special emphasis on the fact that the person interviewed/photographed can request any information that may reveal his/her identity to be kept confidential.
    - Contact information so that the participant can reach the information collector (and vice versa).
    - Details on how long the information will be used, and how and where it will be kept (stored).
    - Reminders that the participant can cease participating at any time (i.e. withdraw consent), and request that his or her information and images/video be destroyed, whenever feasible. The ability of an

organisation to remove an image from the internet once published is extremely difficult. Although it can remove versions from its own files and websites, and stop using it in their materials, images already published may be replicated across multiple sites which the commissioning organisation has no control over. In addition, it is worth noting that once images are published there is often little to stop others using or altering them in order to give a different meaning than intended.

- Instruct those who take photos for the organisation to ensure enough time is given for the informed consent process, including the explanation of what is being asked for, the risks involved, and allowing enough time for the participant to consider whether they wish to agree.
- Develop specific policies and processes to protect children. These should include:
  - consultation with a child's rights expert to ensure the informed consent policy complies with the organisation's legal obligations in the jurisdictions it works in;
  - establishing a formal relationship with a children's rights expert to provide advice on the production and use of images of vulnerable and abused children;
  - establish a mechanism that would automatically refer image(s) that depict vulnerable or abused children to a children's rights expert if the image(s) is being considered for publication;
  - where practically possible, informed consent should be given by the child and their guardian, and such consent should be recorded in a way so that it can be produced on request to those who publish the photo(s); and
  - a commitment that the best interest of the child is prioritised over all other considerations.

The development and implementation of such a policy presents serious challenges to organisations. Photographers working to such policies would be required to adapt their practice in order to comply with the consent protocols of the commissioning organisation. However, it can be argued that not only would such a situation protect the rights of the participants, but also safeguard the photographer (and commissioning organisation) from legal challenges and accusations of unethical behaviour. As such, it could be reasonably argued that all those who use photographs of vulnerable populations in their work would benefit from not only understanding the concept of informed consent but applying a policy similar to that outlined in the framework above.

NOTES

1. Section 23(2) of the Protection of Children from Sexual Offences Act, 2012 (POCSO) and Section 228A of the Indian Penal Code, 1860 (IPC), prohibit printing and publication of any matter which may disclose the identity of a child or of a person against whom an offence under Sections 376, 376A, 376B, 376C or 376D of IPC is alleged or found to have been committed, unless permitted in accordance with the said provisions (https://wcd.nic.in/sites/default/files/childprotection31072012.pdf).

2. Videos of the conference and a short 'Notes and Outcomes' document are available on the Rights Exposure website (https://www.rightsexposure.org/portfolio).

ACKNOWLEDGEMENTS

I would like to acknowledge the invaluable contribution made to this chapter by Smita Sharma. Smita's experience as a photographer working under very difficult circumstances provided much needed insight into the practicalities of the process of acquiring informed consent in the real world. I would also like to thank her for generously providing one of her photos to illustrate this chapter. Smita Sharma's work can be found at http://www.smitasharma.com

REFERENCES

Alam, S. (2007). The visual representation of developing countries by developmental agencies and the Western media. *Policy & Practice: A Development Education Review*, *5*, 59–65. Retrieved from https://www.developmenteducationreview.com/issue/issue-5/visual-representation-developing-countries-developmental-agencies-and-western-media

Campbell, D. (2014). The myth of compassion fatigue. In L. Kennedy & C. Patrick (Eds.), *The violence of the image* (pp. 97–124). New York, NY: I.B. Tauris.

CONCORD. (2006). *Code of conduct on images and messages*. Retrieved from https://concordeurope.org/wp-content/uploads/2012/09/DEEEP-BOOK-2014-113.pdf

Duckrabbit. (2018, November 13). Retrieved from https://twitter.com/duckrabbitblog/status/1062336267562684418

Gabbert, E. (2018, August 2). Is compassion fatigue inevitable in an age of 24-hour news? *The Guardian*. Retrieved from https://www.theguardian.com/news/2018/aug/02/is-compassion-fatigue-inevitable-in-an-age-of-24-hour-news

Goldberg, S. (2018, April). For decades, our coverage was racist. To rise above our past, we must acknowledge it. *The Race Issue, a special issue of National Geographic*, 4–6.

Imaging Famine. (2005). Retrieved from http://www.imaging-famine.org

Lissner, J. (1981, June 1). Merchants of misery. *New Internationalist*. Retrieved from https://newint.org/features/1981/06/01/merchants-of-misery

Mishra, T. (2018). The great upheaval: Can the digital revolution potentially shift the power dynamic in photography? Retrieved from https://viewbook.com/articles/the-great-upheaval

Pixley, T. (2017, September). The western gaze: On photojournalism and challenging harmful representations. *Photo Voice*. Retrieved from https://photovoice.org/the-western-gaze-on-photojournalism-and-challenging-harmful-representations/

The Ethical Image. (2018). Notes and outcomes. Retrieved from https://view.publitas.com/rights-exposure/the-ethical-image_notes-and-outcomes/page/1

United Nations Convention on the Rights of the Child. (1990). Retrieved from https://www.ohchr.org/en/professionalinterest/pages/crc.aspx

CHAPTER 5

POSSIBILITIES AND CHALLENGES: ISSUES IN ETHICAL FILMMAKING USING *IT STAYS WITH YOU* AS A CASE STUDY

Cahal McLaughlin and Siobhán Wills

ABSTRACT

Two western Europeans produced a film It Stays With You: Use of Force by UN Peacekeepers in Haiti *(59 minutes, 2018) about the uses and results of violence by United Nations peacekeeping troops in Haiti. Here, they offer an opportunity to reflect on and learn from an array of themes that include wealth disparity, resource inequality, language barriers, distant communication, and physical and mental health. Such contrasts of access to resources test the possibilities that participatory practices can offer, and this chapter attempts to draw out what was achieved, and what was not, in such a collaboration, by reflecting on what challenges were faced, how these were addressed, and what lessons were learned.*

Keywords: United Nations peacekeeping; Haiti; use of force; participatory filmmaking; unequal resources

INTRODUCTION

In the documentary film, residents of Bois Neuf, a poor neighbourhood of Port-au-Prince, Haiti, describe the impact of the intensive and lethal use of force by United Nations Peacekeepers during the period 2005–2006. UN raids to arrest

Ethics and Integrity in Visual Research Methods:
Advances in Research Ethics and Integrity, Volume 5, 93–103

ISSN: 2398-6018/doi:10.1108/S2398-601820200000005010

gang leaders led to scores of civilian deaths and many more injuries, with testimonies from survivors including the death of a 10-year old girl, the loss of a breadwinner, and the destruction of fragile homes and jobs. The film also includes the views of medical, legal, and diplomatic experts on the responsibilities of the UN in addressing such traumatic events. The survivors said that, even though it had been 10 years since the UN operations took place, the filmmakers were the first people outside of the community to ask them about what had happened.

When the first democratically elected President of Haiti, Jean-Bertrand Artistide, was overthrown in a USA-supported coup d'état in February 2004, the USA sent in troops. These left in June and were replaced by the UN Stabilisation Mission in Haiti (MINUSTAH) a peacekeeping mission mandated by the UN Security Council under Chapter VII of the UN Charter. Today, many Haitians regard the UN presence as an 'occupying force'.[1]

Wills is a professor of law whose research focusses on international peacekeeping (Wills, 2018). Her research into UN regulations on use of force by peacekeepers indicated inconsistences and gaps in the application and interpretation of the UN's legal obligations with regard to use of deadly force for law enforcement purposes, such as arresting criminals, managing riots, and ensuring security at elections. This lack of clarity has resulted in serious violations of human rights, most notably in Haiti between 2005 and 2006 when UN peacekeepers carried out operations that left scores of civilians, including children, dead.

Wills approached McLaughlin, a professor of film studies, to consider recording testimonies of the survivors of these raids in order to present as evidence to the UN. There is a visceral quality to filmed interviews that cannot be represented in written text – we not only read the words, but see and hear the expressions and performance of the testimony giver. During preparations for the filming, it became clear that a documentary film, based on these interviews, might be the best way to represent their experiences and highlight the issues. As Principle Investigator and Co-Investigator, we won an AHRC grant for the project. The resulting film, *It Stays With You: Use of Force by UN Peacekeepers in Haiti,* has been screened internationally – from the UN Human Rights Council in Geneva to the International Law Association in Sidney (www.itstayswithyou.com).

Participatory practices in filmmaking operate on at least two levels: first, they encourage trust between filmmakers and the 'subjects' of the film, by providing a degree of co-ownership (Frisch, 1990); second, they minimise the risk of re-traumatisation by sharing authorship of the survivor's story – a key consequence of trauma can be the fracturing of one's sense of self, or self-narrative (Papadopolous, 1998). McLaughlin's previous experience in Ireland and South Africa, working with those who have suffered trauma from political violence, has ranged across the spectrum of participatory practices, from day-to-day, face-to-face contact to looser, across-the-oceans contact and online communication (McLaughlin, 2012). The Haiti project has provided the greatest challenge to a fully participatory project, given the extreme differences in distance, language and access to resources between the filmmakers and participants. Dana-Ain Davis makes the point that participatory practice 'embodies fairness, respect, and encourages the broadest possible distribution of power .... The challenge is in transferring the ideal of equity into a lived experience of shared power' (Davis, 2006, p. 233).

WHAT WORKED?

Because there is a spectrum of participatory practices, from being involved in discussions and decision making through to operating the camera and co-directing, this case study refers to a specific situation that was limited by our resources (equipment, time, and finances) and relationships (insiders and outsiders to the community affected, along with social and economic differences). Given these challenges, we were situated towards the thin end of the spectrum, but always attempting to ensure that participation underpinned our decisions when possible.

Survivors, or participants, trusted us enough to provide testimonies of their experiences – from the death of a loved one to loss of property to serious injury – and which ranged in nature of expression from poignant to painful to angry. Expectations of what could be achieved were discussed and agreed in advance of filming. At a meeting of those who might contribute, one participant asked us to bring 'justice' for them – they felt marginalised and forgotten – but we had to manage such expectations by explaining our limited resources and experiences in seeking acknowledgement of previous injustices. We could bring public attention to the issue and their stories, but we could not pursue legal cases on their behalf.

These discussions also included the content and style of the proposed film, and it was agreed to prioritise the participants' stories, with some academic and legal expertise allowed as support evidence. The participatory practices continued up to the point of approval of the final film before public exhibition, although the issues of distance and language and resource differences – we had access to camera kits, transport, drivers, and interpreters – tested such collaboration, as we shall describe later.

It is worth noting that another neighbourhood, Project Drouillard, across the main road from Bois Neuf, declined to work with us on this project. Their reasons included 'we want to portray the positive influence the community group is having on the area' and 'we have moved on from the incidents'. While they were prepared to consider collaborating on a film about their efforts to 'clean up' the area, and this would have made an important film, this was not what we were funded for and both parties agreed not to proceed. The right not to contribute is as important an ethical right as the right to contribute, and our previous experience has confirmed this – in Ireland there are as many individuals and communities who wish not to be filmed as agree to be filmed (McLaughlin, 2012). Our experience of working in television has sometimes led to uncomfortable situations when consent forms are required to be signed *before* filming, and certainly editing, takes place, so depriving the contributor of fully informed consent. The importance of providing consent throughout and especially at the end of the editing process and before publication is an acknowledgment of the way that a filmed interview, for example, can be altered in the editing suite. The clip of the full interview could be edited to remove an aspect of the context that offers a different interpretation of the one intended, or the meaning of the clip could be altered depending on where the clip is situated, for example, as complementary or antagonist to the clip before or after. In another project, a prison officer explained that he had once contributed to a television programme, but 'did not recognise' himself in the final edit.

WHAT WERE THE CHALLENGES?

Haiti is rated as one of the poorest countries in the western hemisphere, and Europe one of the richest regions. Given such contrasts in resources – for example, in access to power and wealth between Europe and the Caribbean, and in language, where Haitian Creole is spoken in Haiti and English is the most common language in Ireland – the methodology of participatory practice was severely tested.

Participants live in the Bois Neuf neighbourhood, one of the poorest in Cité Soleil, near the capital Port-au-Prince, with no running water and limited electricity. We had the equipment from our universities, the finance from an AHRC grant, the expertise in our law and film backgrounds, and the resources, which included time, institutional support and the employing of an interpreter, to make this project successful. By contrast, the participants struggled each day to ensure there was a meal on the table for themselves and their children. The risk of dependency was always present, but, as we became aware, operated across different pathways. While participants depended on our resources and expertise, we depended on their experiences and expertise, as well as the interpreter's mediation of the planning discussions, interview questions, and post-production coordination. Throughout the process, transparency of the production process, from research through to exhibition, was encouraged as much as practically possible. We began by holding meetings to discuss and decide upon the project's aims, methods and processes, and we concluded by holding a private screening before public exhibition to ensure the final film was acceptable to the participants.

In a situation where cultural and societal differences are pronounced, the interpreter becomes central to the project and is key to a successful outcome. The role demands not only good language skills but also skills of communication, experience and/or expertise in the subject area, local knowledge, willingness to listen carefully and respectfully, and to be trusted by participants. Our interpreter possessed all of the above, although the difference in languages and priorities could present problems. Mimi Dominique was not from the community, but had previous experience working in the health centre in the neighbourhood, and on each occasion we visited she was greeted warmly by everyone we met. On one occasion, after we had finished the film and returned for a follow-on project, when a local leader of a neighbourhood gang objected to our presence, we had to cancel filming. 'Everyone leave now', the interpreter said, but would not inform us about the detail of the objection. This marginalised and frustrated us, although we realised that the issue was perhaps sensitive. We were unsuccessful in getting further clarification from him or from Mimi.

PRODUCTION

Our first visits involved identifying and negotiating with those who wished to participate in the film. Dominique was crucial in this regard, since she was known by most of the participants personally from her previous work in the community. After a first meeting with the community leaders, we then organised a further meeting where those wishing to participate attended. This visit was intended for research purposes, but with a camera to hand, we were advised to film immediately, because, given the precarious nature of life in the neighbourhood, there was no guarantee that the participants would be around for our next visit. The interviews were conducted with Wills interviewing, Dominique interpreting, and McLaughlin on camera and sound. Given the small size of most dwellings, this made for a compact set-up, with several occasions when children moved across the front of the camera, but, given the nature of the environment, it was neither desirable nor necessary to intervene, for example, by asking them to stay out of shot. We visited the neighbourhood on four occasions – twice to record, once to have the participants check on the final film before screening publicly, and once to record some individual pieces for our website.

The previously mentioned example of having to leave the neighbourhood abruptly was the most extreme example of the unpredictability inherent in the project, which itself reflected the daily precariousness of living for all of our participants. On one other occasion, we had intended to spend several days recording general footage of the participants and their everyday lives (walking, working, laundry, and meetings). However, we were informed by the community leader, who had been central to early negotiations and accompanied us at all times, that we had only one hour, because of his concern for our and the participants' safety. This happened during a presidential election when it was expected that there might be community disruption, and possibly violence, because of the contested nature of the election process, which was regarded by most locals as corrupt.[2] Since the safety of the participants was a priority for us also, as well honouring the advice of the community leader, we were happy, if reluctant, to comply. One aim of filmmaking is to provide visual evidence of the narrative, so that audiences have a stronger sense of 'being there', of immersion in the story. However, such a desire has to be balanced by the needs and experiences of the community, and of the conditions prevalent when filming. On this occasion, the priority of the community, as interpreted by the community leader, was paramount.

POST PRODUCTION

While filming in Bois Neuf and neighbouring Port-au-Prince, we at least had direct contact, albeit mediated by the interpreter, with the participants. When we were editing the film, known as post-production, this contact was lost. Unlike previous experiences in South Africa, where the Human Rights Media Centre (producers of the films, *We Were There I* and *II*), acted as mediators with the participants from the organisation Khulumani Western Cape, there was no such organisation in Haiti. Here, the interpreter was our only contact point. This was based on a relationship dictated by employment terms, and, while she occasionally visited Bois Neuf, this was by no means a regular occurrence. Thus, contact was very limited and we relied even more heavily on the interpreter for communication than before.

Given the limited digital resources in Bois Neuf, we were unable to send work-in-progress to the participants, but instead presented a draft edit on only one occasion. The response was positive; the only change required was the need to replace subtitling of English language interviews with dubbing because some of the survivors had difficulty with literacy. We also showed the complete film to the participants before the premiere at the Soros Foundation-funded centre in Port-au-Prince, FOKAL, with the facility to make basic edits if there any requests. On both occasions, the participants did not raise any issues that needed to be addressed in the editing, which suggested that we had made a film that they were satisfied with. However, given limited access to resources, including time to take the film home to view and consider, limited contact with the filmmakers and therefore limited communication and discussion, and limited expertise in filmmaking, the lack of amendments may be put down, at least partially, to the difficult conditions under which we were applying participatory practices. It could also be that our early discussions set the groundwork for a process that we adhered to.

The degree of participatory practice was tested by this project. Ideally, we would have liked the resources to spend more time with the community to deepen the participatory relationship, but, because of our limited resources, the degree of participation was constrained. Given that research councils provide funding for teaching relief and expenses, there is always a time limit to a funded project. Projected costs are estimated from plans that, however, may be subject to unexpected change after the grant is awarded. With more funding, more visits could have been arranged, but given the political uncertainty of Haiti (which continues to this day),[3] it would be difficult to estimate in advance what would be a guaranteed schedule of activity.

To compensate in some way for the above challenges, we ensured that we provided as much opportunity as possible for transparency, so that participants were making informed consent. The limitations of our time and the disparities of resources were included in the information we provided in our discussions with the participants at each stage of our relationship (we visited the neighbourhood on five occasions), so that informed decisions could be taken by them, balancing what was desired by both filmmakers and participants, with what was achievable and finally what was acceptable to all parties in the conditions we faced.

EXHIBITION

While managing expectations around our inability to offer 'justice' to the participants, we had, and have, a responsibility to ensure that their voices are heard and listened to, especially in the context where some beneficial action might follow. This political commitment was threaded throughout with a personal commitment; it is difficult not to increase one's resolve when witnessing the father of a teenage girl break down in tears when remembering how she died, nor the anger of a mother, whose child died four months after being shot by heavily armed soldiers, who offered no medical support after the raids.[4]

Initially, we hoped that some form of acknowledgment by the UN might be forthcoming – which could range from an independent and transparent investigation in the deaths and injuries (which we have called for), through to an apology for the loss of life, injuries and destruction of homes caused by the raids. Engagement with the UN has been difficult, frustrating and a task that two academics do not have the resources to challenge effectively. The imbalance of forces that existed between the participants and the filmmakers has been replicated between the filmmakers and the UN. We can only guess the alienation, hurt, and humiliation that the participants feel in being attacked and violated, then misrepresented as gangsters (Hallward, 2007, p. 282), and finally told to go away by the UN.[5]

We realise the need to invest as much time and finance in exhibition and distribution of the film as in research and production, but the resources needed to challenge authority, especially as large and closed as the UN, requires resources that a relatively small grant could not sustain. Further, we have to consider that there are so many issues of concern in the world, and so many communications about them, that we are facing an uphill battle just to be heard. Because the film was intended to have an effect, that is, to get the UN to take notice, at best to hold an investigation and at least to acknowledge the innocence of the victims of their raids, we regarded the film as part of a bigger project, as a means to an end. The limited response from the UN is undoubtedly disappointing, but has not lessened our wish to have the testimonies of the survivors acknowledged publicly.

The film has had a mixed exhibition journey. Using some online festival sites such as Film Freeway, we have been surprised at the lack of interest by human rights and documentary film festivals. Of course, we need to be modest in terms of what we were trying to achieve aesthetically. We operated participatory practices in a challenging environment and spent only four short visits in Haiti with limited access to resources there. Nonetheless the rejection rate has been higher than we first anticipated. We believe that there may be three reasons to consider: Haiti is not a priority in terms of human rights, and if it was, the issues of cholera and sexual violence perpetrated by the UN have taken precedence (Hallward, 2007). Second, challenging the UN, with its reputation for peace-keeping, mediation, refugee support, education and human rights, runs against the grain of most people's expectations of its role – to accuse it of reckless lethal violence does just not compute. Finally, the film does not follow the usual narrative structure of following one or two people on a journey, meeting obstacles and overcoming them. Participatory practices demand working with the participants' ideas of what makes a good film – in this case the testimonies of the survivors dominate the story and the structure. It could be argued that there are 'types' of films that are preferred by various exhibition platforms, for example, broadcast television generally prefers, although not exclusively, presenter-led documentaries, while film festivals prefer, again not exclusively, story arcs, where individuals or communities are followed over a period of time. Participatory filmmaking for research purposes could adapt one of these models, but would run up against the 'authorial voice' in the first instance and limited resources in the second instance.

While the main aim of the film was to have survivors' testimonies heard, and ideally acted upon by the UN, the secondary aim of providing a public platform for voices that were ignored or silenced was discussed and agreed between the researchers and participants at the preliminary meetings. In this project, while the first aim has not yet been successful, the second aim has had reasonable success. Disappointment has been leavened by participants' voices reaching international platforms.

The film has been shown internationally, from Australia to Japan to Canada to Uruguay and the USA. While most have been academic venues, which can have an impact (if less noticeable), some have been professional, for instance, at the International Law Association Conference in Australia. The film festivals have included: the Caribbean Tales Film Festival in Toronto; Respect Human Rights Film Festival, Belfast (where the film was awarded Best Director); the Amnesty International Film Festival, France; and the Tenemos Que Ver Human Rights Film Festival, Montevideo. Probably the two most significant screenings were the premieres in Port-au-Prince and at the Human Rights Council side session in Geneva in July 2018. At both FOKAL and the University of Port-au-Prince, two participants sat on the post screening discussion panel in order to ensure their sharing responsibility for the film and its impact. The Geneva screening was hosted by the International Federation of Human Rights and the discussion included Rosie Auguste, the Haiti spokesperson for the National Human Rights Defence Network. Consequent to these screenings, the Haitian-based UN human rights officers arranged for participants in the film to be interviewed. Those who were available were invited to the UN offices in Port-au-Prince, rather than a visit arranged to their damaged homes. Several months later, we were informed that no further action would be taken and we should communicate directly with UN headquarters in New York. Despite several emails, we have had no response to date.

FOLLOW UP RESPONSIBILITIES

After the film was produced and exhibited, we had to acknowledge the lack of response on the lobbying front. We decided to revisit Bois Neuf to offer the opportunity for participants to each make a short film about an issue they felt was important – we brought small, easy-to-use cameras, with the intention of lending them to the participants for several days and then organise local hands-on editing sessions. This was the occasion referred to previously when we were told to leave abruptly, so instead we all moved to a nearby empty school and the participants decided to there-and-then make short to-camera pieces. In order to restore some agency to the participants, we invited them to select their own theme, for example, a digital memory of their loved one, which we felt would provide a lasting memorial, or maybe an address to the UN. To our slight surprise they unanimously decided to directly address the UN, which reflected their on-going grief, frustration and anger at the damage and silencing, even 10 years later. These appeals to justice and restitution can be seen at https://itstayswithyou.com/updates/. We made these short films with the assistance of two young Haitian filmmakers, who were brought in for a workshop to teach filming and editing to the participants, but, given the circumstances of having to leave abruptly, instead they provided the technical assistance to produce the to-camera pieces and took still photographs that we later used on the website; for these short films, it was decided that Jeanty Junior Augustin and Pierre Moise have director credits in order to acknowledge their crucial input at a difficult time. We sent these individual testimonies to the UN Secretary General and Under Secretary-General for Peacekeeping, as part of our on-going commitment to bring the survivors' allegations and request for investigation to the attention of the UN, but to date have had no response.

The project has confirmed for us that it is important to allow space for audiences' emotional reactions to such painful stories, before engaging in law and policy debates on how to address the problems presented. An important way to communicate with film audiences, on any subject, is to engage with the emotions that lie just under or even on the surface of such testimony-giving, and also be aware of the risk of re-traumatising audiences who may have suffered their own painful experiences. Our responsibility to audiences therefore is to provide advanced information about the difficult nature of the subject, which we ensure in our promotion material and at the beginning of the film – using text to set out the issues and context.[6] When we premiered in Port-au-Prince, space was given to frustration and anger at the on-going political instability in the country; in Toronto, the Haitian diaspora shared the hurt and pain of the participants; at Harvard, the chair asked for a moment of reflection after the screening followed by a sharing of personal responses from the audience before the panel of experts began their discussion – in order not to lose the sense of community engagement.

CONCLUSION

Participatory practices in filmmaking are severely tested when collaborating across cultural, social, economic, and political barriers, which include resource disparity, language differences, distance, and divergent motivations (Figs. 1 and 2). In the case of *It Stays With You,* early discussions attempted to manage expectations for both parties – filmmakers and survivors of violence – in order to collaborate on the production of a documentary film that prioritised the experiences of those who lost loved ones, were severely injured, and/or lost their property. Our limited successes include participant engagement with the recording of the film, satisfaction with the final edit, and participation in post-screening discussions in Port-au-Prince. Such engagement was intended to provide a platform for those affected to have their voices heard; on one occasion during filming, Joseph Pierre-Louis, who lost his house, became homeless, and has since passed away, raised his arms above his head and said, 'Since you have come, I have grown three foot taller'. Our limited successes in having these voices heard include screenings and discussions at some relevant university departments, for example, New York, Tokyo, Sydney and Berlin, key Caribbean and human rights festivals, for example, Toronto, Belfast, Montevideo and Nice, as well

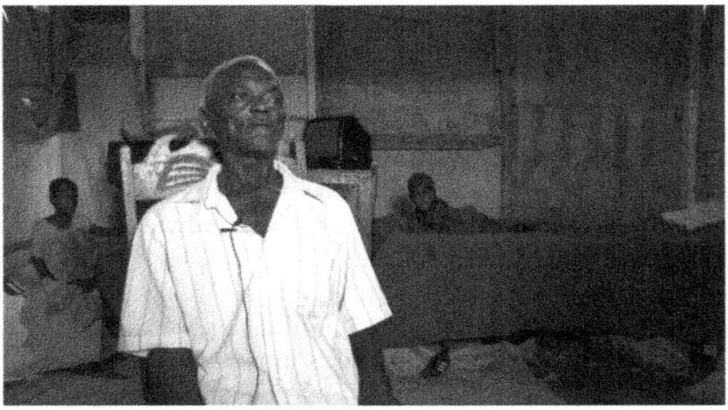

Fig. 1. Victor Jean. Still taken from the film *It Stays With You.*

Fig. 2. Evelyn Myrtil. Still taken from the film *It Stays With You.*

as an important screening and discussion at the UN Human Rights Council session side event in Geneva. We have seen this film dozens of times, but tears and anger still well up when we hear and see Victor Jean cry at the loss of his young daughter, or when Modeline remembers the death of her father. We have realised that our initial hope for recognition by the UN has been tempered, but is not extinguished.

NOTES

1. For a detailed history of Haiti, from the slave revolt that led to the first Black Republic up to the arrival of UN peacekeeping troops, see Peter Hallward (2007).
2. An example of such criticism can be seen in 'Haiti election 2006: False hope, barbarism, primitivism, mob violence' (2006).
3. See further 'Haiti protests: Violent clashes in Port-au-Prince' (2017).
4. As I write this, my own eyes well up with pain and anger at these memories.
5. One participant was told this when she went to a UN military camp to complain about the death of her husband. 'I gave up', she tells us.
6. More information about the project can be found at www.itstayswithyou.com.

REFERENCES

Davis, D. A. (2006). Knowledge in the service of a vision: Politically engaged anthropology. In V. Sandord & A. Angel-Ajahi (Eds.), *Engaged observer: Anthropology*, advocacy and activism. New Jersey, NJ: Rutgers University Press.

Frisch, M. (1990). *Essays on the craft and meaning of oral and public history*. New York, NY: Suny Press.

Haiti Election 2006: False hope, barbarism, primitivism, mob violence. (2006, February 13). Retrieved from http://www.wehaitians.com/haiti%202%20election%202006.html

Haiti protests: Violent clashes in Port-au-Prince. (2017, September 13). *BBC News*. Retrieved from https://www.bbc.co.uk/news/av/world-us-canada-41251893/haiti-protests-violent-clashes-in-port-au-prince

Hallward, P. (2007). *Damning the flood: Haiti and the politics of containment*. London: Verso.

McLaughlin, C. (2012). *Recording memories from political violence: A filmmaker's* journey. Bristol: Intellect.

Papadopolous, R. (1998). Destructiveness, atrocities, and healing: Epistemological and clinical reflections. *Journal of Analytical Psychology*, *43*(4), 455–477.

Screenings. It stays with you: Use of force by UN peacekeepers in Haiti. Retrieved from https://itstayswithyou.com/screenings/

Wills, S. (2018). Use of deadly force by peacekeepers operating outside of armed conflict situations: What laws apply. *Human Rights Quarterly*, *40*(3), 663–702.

CHAPTER 6

NAVIGATING THE NECESSARY RISKS AND EMERGENT ETHICS OF USING VISUAL METHODS WITH MARGINALISED PEOPLE

Jacqueline Shaw

ABSTRACT

The global call to 'leave-no-one behind' cannot be achieved without tacking the intractable social issues faced by the most excluded people. There is increasing interest in using visual methodologies for participatory research in contexts of marginalisation, because they offer the potential to generate knowledge from people's lived experience, which can reveal subjective, emotional, and contextual aspects missed by other methods; alongside the means for action through showing outputs to external audiences. The challenge is that the perspectives of those in highly inequitable and unaccountable contexts are – by definition – rarely articulated and often neglected. The author thus begins by assuming that there are unavoidable tensions in using visual methods; between perpetuating marginalisation by inaction, which is ethically questionable; and the necessary risks in bringing unheard views to public attention. Many experienced practitioners have called for a situated approach to visual methods ethics (Clark, Prosser, & Wiles, 2010; Gubrium, Hill, & Flicker 2014; Shaw, 2016). What is less clear is what this means for those wanting to apply this practically. In this chapter, the author addresses this gap through the exemplar of participatory video with marginalised groups. Drawing on cases from Kenya, India, Egypt, and South Africa, the author contributes a range of tried-and-tested strategies for navigating the biggest concerns such as informing consent; and the tensions

Ethics and Integrity in Visual Research Methods:
Advances in Research Ethics and Integrity, Volume 5, 105–130

ISSN: 2398-6018/doi:10.1108/S2398-601820200000005011

between respecting autonomy and building inclusion, and between anonymity and supporting participant's expressive agency. Through this, the author provides a resource for researchers, including prompts for critical reflection about how to generate solutions to visual ethical dilemmas in context.

Keywords: Marginalisation; situated ethics; visual methods; participatory video; inclusion; group process

1. INTRODUCTION

For social research to address complex real-world problems, it needs to generate *phronesis* or Aristotle's practical wisdom (Flyvbjerg, 2001). A phronetic approach focusses on constructing contextual understanding through interaction with people affected by an issue, and this resources action towards improved solutions (Tracy, 2013, p. 4). The global call to 'leave-no-one behind' (Kabeer, 2016, p. 5) is unlikely to be achieved without collaborating with the most marginalised groups to build this kind of grounded knowledge, and using the insights to influence responses from external agencies. Participatory researchers are enthusiastic about the promise of visual methodologies, such as digital storytelling and participatory video, in contributing to these research and communication aims (Gubrium, Harper, & Otanez, 2015; Mitchell, De Lange, & Moletsane, 2017, p. 4; Shaw, 2015, p. 4). Although it is not a singular approach, broadly participatory video is a facilitated group process, which involves participants recording themselves and the world around them, reflecting on their situation together, and producing video stories or messages in order to generate further dialogue across communities or with external agencies (Shaw, 2017a, p. 5). Digital storytelling involves participants crafting individual stories in a group context, and then recording their personal narratives with accompanying visuals and music in a digital format (Lewin, 2011, p. 55). These approaches provide an accessible way for people to explore and share experiences. They are appropriate for group inquiry processes because they are essentially relational: they prompt group members to speak about their lives, and listen to each other; together they make sense of their realities through reflective dialogue; and later, pathways to influence are an inherent possibility through a group creating and then showing their visual narratives to external audiences. The key ethical dilemma is that, because the perspectives of excluded people in highly inequitable and unaccountable contexts are – by definition – rarely articulated and often neglected, there is concurrent danger in bringing them to public attention. In practice, navigating ethical dilemmas is essential as a process unfolds, with ethical relations at the heart of socially engaged research excellence.

The contemporary resurgence of participatory visual and performative methods to generate missing knowledge on intractable social issues has occurred in parallel with increased ethical regulation in the social sciences and allied disciplines (Wiles et al., 2008, p. 10; Wiles & Body, 2013, p. 1). It is not surprising that critical debate has been provoked about the ethics and integrity of visual methods, given the challenges posed to research ethics norms such as respondent and geographical anonymity, and the difficulties of informing consent given participant's desire to be heard in the sharing age (Clark, Prosser, & Wiles, 2010, pp. 85–87). Visual researcher-practitioners[1] motivated by justice principles, and the potential for these methods to empower participants through building agency and providing communication possibilities, favour showing visual materials in their entirety, if it is appropriate to show them at all (Wiles et al., 2008, p. 22). However, this does not mean anything goes. It is clearly not ethical to encourage participants on short-term projects to reveal deep feelings on sensitive issues, and disseminate exposing visual materials with insufficient thought or clarity about audience, platforms, access, or longevity (Clark et al., 2010; Prosser & Loxley, 2008; Wiles, Crow, Charles, & Heath, 2007b). This is the case whether or not participants are recognisable through models, drawings, photos, digital stories, video, or accompanying text. Many experienced practitioners have called for a *situated* approach to visual methods ethics (Clark et al., 2010, p. 83; Gubrium, Hill, & Flicker, 2014, p. 1606; Shaw, 2016, p. 421), which takes into account the particular circumstances of who, what, why, and where. What is less clear is what this means practically for people wanting to apply visual methodologies with marginalised groups, especially given the diversity of people and contexts encompassed by this generalising terminology.

Currently, I am a research fellow specialising in the use of visual processes to drive and mediate participatory action research (PAR) in community, development, and health contexts. Situated ethics are specific, so I distinguish between the one-off/short-term use of visual approaches in participatory research interactions, which is not the focus here, and the progressive visual methodologies introduced above. In this chapter, I address the gap in practice knowledge about these longer-term research processes, by concentrating predominately on the exemplar of participatory video. To do this, I draw on more than 35 years' experience using participatory video with a diverse range of disadvantaged groups, including women, people with disabilities, homeless, and migrating people, elderly and young people, and others living in insecure contexts or discriminated against due to factors such as gender, ethnicity, caste, class, sexuality, health status, or geography. I do not underestimate the considerable challenges involved in using visual approaches to transform iniquitous dynamics, and I have developed my approach to participatory video (Shaw, 1986; Shaw & Robertson, 1997), through research to build critically nuanced understanding about the practical realities (Shaw, 2012a, 2012b, 2016). Applying the knowledge generated on what enables and constrains the possibilities of participatory video, given the inherent risks and challenges, has resulted in the extended participatory video methodology I currently favour (Shaw, 2015, 2017a, 2021 forthcoming). In this chapter, I consider the application of this longer-term participatory video approach during the *Participate* research initiative[2] (2012–2018), which first brought perspectives on the reality of marginalisation into the global post-2015 debate (Burns, Howard, Lopez-Franco, Shahrokh, & Wheeler, 2013) then developed knowledge on participatory accountability mechanisms (Burns, Ikita, Lopez-Franco, & Shahrokh, 2015), and more recently explored how to build inclusion of the most marginalised people in the context of Sustainable Development Goals (SDGs) implementation (Howard,

Lopez-Franco, & Shaw, 2018, p. 6). I use this research context to illustrate the strategies I have evolved, from the early 1980s onwards, for ethical participatory video practice in participatory research and community development projects.

I begin, in Section 2, by framing the inherent risks of research collaborations with marginalised groups, and in Section 3, I outline current thinking on how to respond to the ethical challenges presented by visual methods. In Section 4, I present the extended participatory video approach as the basis for ethical practice, and I define two key tensions faced when navigating between avoiding harm and doing good through facilitated participatory video processes: supporting participants' autonomy when building the inclusion of marginalised groups; and respecting participant's choices between anonymity and public communication, whilst evolving and surfacing knowledge from the margins.

In Section 5, I unpack the general strategies that I have used successfully as a foundation for ethical practice, in relationship to specific examples from the *Participate* research projects. Then, in Section 6, I draw again on *Participate* experiences, as well as other pertinent examples, to raise the need to adapt participatory video processes according to the particular group the starting point of individuals, the contextual challenges and the sensitivity of the issue. I finish in Section 7 with a number of questions to prompt researchers and practitioners planning their own participatory video projects.

In this chapter, I contribute a range of practical solutions to the key ethical and integrity concerns of visual methodologies, through considering the exemplar of participatory video. This includes many tried-and-tested strategies: such as the use of progressive processes to support participant's choices and inform consent; the involvement of two practitioners to provide both one-to-one support and group-orientated facilitation; the practice of group members being subjects of their own videos; the clear separation of video recorded to prompt group refection in *safe* confidential space, from external video communication; the importance of raising awareness that no video materials will be shown outside the group unless or until participants chose to; the need for sufficient time and cycles of the process to enable capacity-building, agenda development and deeper critical insight; the practice of showing videos in progressively diversifying social spaces; and the careful contextualisation of video materials for external audience including participant's interpretations of the meanings. This practice model provides a resource for researchers and programme leaders wanting to apply participatory video methodologies ethically, by assisting them in anticipating the likely risks, by prompting them to think critically about what needs to be addressed, and by providing specific ways to avoid the worst difficulties.

2. THE INHERENT AND NECESSARY RISKS OF RESEARCH COLLABORATIONS WITH MARGINALISED PEOPLE

Marginalised groups and communities are those that face systemic disadvantage and discrimination in accessing services, interacting with social institutions, and influencing political processes. Marginalisation is due to the power imbalances that arise from socially constructed identities such as gender, sexuality, race, caste, ethnicity, class, ability, and age, and the most marginalised people often experience intersecting inequalities: they have more than one disadvantaged social identity, which is often compounded by geographic, economic, or political factors (Howard et al., 2018, p. 8). Marginalisation is less about people being overtly pushed to the social margins, but more often a consequence of the lack of action to address the discriminatory attitudes, damaging social norms, and inequitable dynamics that maintain it.

There is a compelling rationale for engaging marginalised groups in participatory research collaborations. Pressing and intransigent social issues are unlikely to be solved without understanding the realities from the perspectives of those facing them. There is ample evidence that imposing solutions top-down, without contextual knowledge often generates unintended negative consequences (Shaw, 2015, p. 2), and downward cycles of poverty for the worst off (Burns et al., 2013, p. 9). It is thus crucial to learn from them to avoid investments failing. Furthermore, the global SDG call to leave-no-one behind will not be achieved without considering the effects of intersecting inequalities (Kabeer, 2010, p. 10; 2016, p. 5), because the particular ways in which marginalisation is perpetuated are overlooked.

PAR (Kemmis & McTaggart, 2005) is a recognised methodology for tackling stubborn social problems, because it focusses not only on exploring the barriers to positive change in complex contexts, but also on generating practical knowledge on solution pathways through cycles of action followed by reflective learning and adaption. The most recent *Participate* research has shown that this can build the insights needed for better governance, and also that PAR processes are in themselves ways to tackle marginalisation (Howard et al., 2018, p. 35). For example, our research partners in extremely inequitable contexts in Egypt, Ghana, India, South Africa, and Uganda demonstrated how local power dynamics can shift through increasing participants' sense of can-do, developing a group's knowledge, rights awareness, and agency; and building influence through working together. This can then provide the necessary leverage to foster accountable relations and hold decision-makers to commitments (Burns et al., 2015, p. 6). However, there is an inherent risk in this endeavour because it is about supporting unheard groups in speaking truth to power.

In this chapter, I proceed from the assumption that research collaborations with marginalised groups happen in unavoidably tensioned contexts because there are conflicting motivations and expectations from different stakeholders. Completely avoiding any possibility of negative consequences following the ethical imperative to *do no harm* means inaction, which is ethically questionable. Yet, it is important to acknowledge that risk-taking is necessary due to the intention to shift inequitable power dynamics. To clarify, research participants may be those such as people with disabilities, who are automatically flagged up as high risk by research ethics check lists. In addition, research interaction may focus on sensitive issues or bring up painful feelings, not least those caused by being discriminated against. However, avoiding research with people at the social margins is also unethical, because their needs and views continue to be publically unknown or unheeded, which contravenes the research ethics principle of justice.

If participatory methods are to not only describe the world, but also *do good* by forging new social possibilities, recognising and responding to tensions is fundamental to ethical practice amongst the messiness (Law, 2004). Researcher-practitioners must face squarely up to the impossibility of completely avoiding risk. Instead, it is crucial to develop awareness of the likely challenges, and how best to not only mitigate risks before a project starts, but also to negotiate them in context by proceeding reflexively as dilemmas arise. This calls into question a pre-project audit approach to research ethics, which assume fixed and asymmetric power relations between researcher and researched, and does not adequately consider the evolving dynamics of collaborative working or co-construction; compounded when institutional ethics review board do not comprehend shifting power relations as a research purpose (Manzo & Brightbill, 2007). Visual approaches add another layer of intensifying complexity, so I next ground my discussions in current thinking on the need for a situated approach to visual ethics.

3. TOWARDS A SITUATED, RELATIONAL, AND EMERGENT ETHICS FOR VISUAL METHODOLOGIES

There is considerable debate about the ethics of visual research methods due to the clear challenge they present to common ethical principles (Clark et al., 2010; Wiles et al., 2008). Rose Wiles (2013, p. 12) summarises the ethical frameworks that can guide qualitative researchers in making ethical decisions alongside their moral orientation, and the legal regulation, professional guidelines, and disciplinary norms. The principalist framework (Beauchamp & Childress, 2001) that informs institutional ethics reviews includes the four pillars of ethical practice: respecting autonomy, avoiding harm, aiming to do good, and justice, and the key practical responses of voluntary participation, informed consent, right of withdrawal, confidentiality/anonymity, treating people fairly, and sharing the benefits of research (Wiles, 2013, p. 6). Constructing knowledge through collaboration challenges the traditional approaches to research authorship and ownership. However, the misalignment between the principalist framework and visual methods becomes most acute in the practice balances between respecting participant's autonomous choices about public communication versus anonymity, which involves supporting them to make informed choices about what to reveal to which audiences and where. This is especially given the ease with which images can be distributed, and the limited control over their use once they are online.

The problems of anonymising visual materials through removing names or using pseudonyms often counters participant's wishes (Wiles et al., 2008, p. 16), and can involve an abuse of researcher's decision-making power in silencing people or making their contributions invisible, which is incompatible with co-construction principles (as with text based reporting) (Shaw, 2016; Walker, Schratz, & Egg, 2008). Obscuring or blurring faces can make a mockery of the use of visuals (Prosser & Loxley, 2008; Wiles et al., 2008, p. 22), as one aim is to provide a route for marginalised groups to be seen and heard. Worse, blanking out eyes is de-humanising, disrespectful, and ineffective, and pixilation makes people look like criminals (Banks, 2001). In response, many visual researchers have asserted that participants have a right to be seen (Clark et al., 2010; Wiles et al., 2008), which is particularly important for those currently less visible such as people with intellectual disabilities (Boxall & Ralph, 2009, p. 45).

In terms of participatory video, there is clear possibility of increasing participant's influence through showing videos (Shaw, 2017b). Visuals materials resulting from participatory projects are grounded in the collaborative relations established (Pink, 2007; Rose, 2007), and the intention is participants decide who to communicate with and what to say (Gubrium et al., 2015; Mitchell et al., 2017). In this way, participatory video can involve a group making a video to support their own agendas. However, in contexts of marginalisation, there is a key tension between protecting vulnerable participants from inappropriate exposure or backlash, which can be paternalistic, and supporting them to articulate their perspectives in public space, when the consequences cannot be controlled. This reflects the ethical dilemmas with other visual methods (Wiles et al., 2008).

Relational ethical frameworks such as the ethics of care (Gilligan, 2008) are generally thought more applicable than the principalist framework to participatory visual processes (Banks, 2001, Wiles et al., 2008), and many authors have proposed a situated and relational approach to ethical decision-making (Clark et al., 2010; Gubrium et al., 2014; Shaw, 2016; Simons & Usher, 2000). This is reflected in participatory video literature where, for example, Capstick (2012, p. 272) suggested a *situated* approach was needed to treat people with physical, sensory or cognitive disabilities *fairly* on a project with people with early-onset dementia. Participatory video practice must be contextualised as people have different needs and wants, and this *is* a relational process played out in highly personalised interactions between the practitioners and group members. However, there is insufficient knowledge of the inter-subjective practice aspects or *soft skills*.

Furthermore, contextual choices may not be straightforward. Whilst, there are resources to help researchers think through the issues (Papademas, 2004), there are questions about whether people can ever know what showing images will be like before the event (Prosser, 2000, p. 118), and particularly how to inform consent in research collaborations with children and other participants who may be more vulnerable due to literacy and cognitive capacities, the circumstances or the research issue, or when consent is not culturally understood as the researcher intends (Pink, 2007). These dilemmas are highly pertinent to research in contexts of marginality (Smith, 2008). My extensive experience and research suggests participatory video ethics should be approached as an ongoing emergent matter (Shaw, 2016). If transforming dynamics productively over time is more important than the videos produced, then contextual adaptation cannot be resolved at the start. Specifically, what happens should develop responsively as processes unfold. Following, Broyles, Tate, and Happ's (2008 in Clark et al., 2010, p. 81) assertion that filming ethics can be negotiated if the territory is clearly defined, I next turn to participatory video practice as a specific exemplar to consider what enables and constrains ethical navigation of the inherent tensions in context.

4. NAVIGATING BETWEEN *AVOIDING HARM* AND *DOING GOOD* USING EXTENDED PARTICIPATORY VIDEO PROCESSES

In Section 2, I explained the inherent risks of participatory research with the most marginalised groups, which are an unavoidable component of addressing intractable inequality to build influence from the social margins. I have defined the overall ethical balance between *avoiding harm* and *doing good*, which it is necessary to navigate during participatory visual processes in these contexts. My previous research (Shaw, 2012b, 2016) identified that the most acute ethics issues in participatory video occur on short-term production projects, where group film-making is predominately perceived as data generation. In this application, participants learn basic production skills, and are supported to make a collaborative video over a short period (e.g. a week or so). Following production, the video is shown to an audience to stimulate discussion. Perceiving the video as the main outcome can constrain potential benefits such as capacity-building, inclusive dynamics, group ownership or collective agency (Shaw, 2017a, p. 5). Production needs often take precedence, with activities to facilitate deeper critical reflection and analysis either truncated or not occurring. Well-recognised ethical risks, such as inappropriate exposure for vulnerable groups (Howard et al., p. 36; Shaw, 2015, p. 8), are also amplified due to brief engagement timescale. Duty-bearers and decision-makers may watch videos during consultation meetings, but this often functions to end rather than begin longer-term citizen-state partnerships (Shaw, 2015, p. 5; 2017a, p. 5). This can *do harm* by exacerbating feelings of exclusion when nothing transpires for those involved.

In response to the risks, I first approach participatory video as an essentially relational practice rather than being technology or production-focussed (Shaw, 2017a, p. 8). This means to videoing, playback and editing activities serve different interactional purposes at the progressive stages. This separates group film-making from the other social aims, as summarised in Table 1. This defines four key participatory video phases, and the rationale for how videoing and playback is applied.

During the group-building and internally-focussed group exploration phases, short videoing and playback exercises are used to build trust, confidence, and inclusive dynamics; and to explore shared concerns through videoing activities in the locale. Whilst this can generate research insight, this early material is not shown outside the group. Clear separation of internally focussed videoing to stimulate research discussion in safe, confidential space, from video's later use to mediate external communication activity is a principle strategy used to mitigate risks. It avoids the danger of inappropriate exposure, before participants have had enough experience to inform their choices about what to reveal to others. Then, once participants have had time to rehearse expressing themselves in private, and formed opinions or deepened their understanding, they are more ready to decide what to convey to whom. Only then, at the externally focussed production phase, do participants make video materials to stimulate dialogue with other audiences.

Table 1. Key Phases of Participatory Video (Adapted from Shaw, 2012b).

Key Participatory Video Phases	PAR Purpose	Video/Playback Activities
Group forming and building	• Motivating engagement – space for something new • Increasing individual communication confidence and capacities • Establishing inclusive dynamics	• Video is an exciting opportunity • Short video recording and playback exercises progressively build agency and sense of *can-do* • Structured role rotation supports inclusive dynamics
Group exploration and reflection	• Co-operative inquiry – dialogue on participants lives and issues in *safe space* • Time for agenda emergence • Deepening and re-framing understanding through critical reflection	• Video exercises provide exploration rationale • Practicing self-expression builds expressive capacities • Group reflection on playback helps people form and articulate opinions • Progressive narrative work can deepen discussions and criticality
Collaborative production	• Exercising collective agency • Exploratory inquiry on key issues • 'Becoming' social actors in semi-public space	• Video provides the means to ask questions and gather views • Recordings can reveal subjective aspects of experiences • Videoing action provides a way to pursue agendas by showing and telling
Video-mediated exchange	• Involving community stakeholders through further production/playback action • Performing influence through showing videos in external forums – in-between space	• Video can provide the engagement hook to initiate exchange • Having videos to show can position participants more influentially • Videos can connect audiences with people's realities and provoke, illicit, or compel response

Recent critical discourse on participatory visual processes emphasises the ethical imperative to apply recording/making and playback activities as iteratively evolving processes during longer-term engagement (Mistry et al., 2016; Shahrokh & Wheeler, 2014; Shaw, 2015). As with situated ethics, what actually happens should develop according to particular circumstances, so what these progressive processes involve is less clear to outsiders. This makes it hard for others to adopt them. Whilst Table 1 functions as an analytical device to elucidate broad phases of the participatory video process, it only encompasses one cycle of videoing activity rather than the iterative cycles that are proposed. However, an extended participatory video process is anticipated to involve repeated cycles of collaborative video-making action, followed by reflective dialogue after playback in progressively diversifying spaces as a project evolves. A group may first create video materials to show to their peers, then to instigate dialogue in the wider community, and then later to external audiences such as service providers or decision-makers at different social levels. This provides the opportunity for participants to hone their public communication skills firstly with audiences likely to be sympathetic, and later in less supportive forums. This navigates the risks of exposure or negative responses more ethically. Ideas are also developed through deliberation with different stakeholders, and this deepens and strengthens research insight, as group videos are not the end of an engagement, but used as a way of opening various new conversations.

Despite the success of these strategies as ethical starting points, re-framing participatory video as the means to drive social interaction, separating internally and externally focussed uses of video, and supporting iteratively progressing cycles of videoing action and reflective dialogue does not make it unproblematic. Rather it makes it apparent why tensions are intrinsic to the practice territory. Indeed, my research shows that the possibilities of participatory video are intrinsically linked to the tensions at each stage (Shaw, 2012a, 2012b, p. 137). Developing this idea, Table 2 summarises the most pertinent balances in relationship to the overall ethical conundrums I have identified in video projects with marginalised groups.

Next, I discuss what can be done in response to these practice balances.

Table 2. Participatory Video Ethical Tensions and Practice Balances (Adapted from Shaw, 2012b).

Main Ethical Tension	Practice Balances
Supporting decision-making autonomy whilst building inclusive dynamics	• Tackling or perpetuating marginalisation • Practitioners' action to establish/maintain inclusive dynamics versus participant choice • Video as enabler versus balance of individual, group, and wider community needs and risks
Anonymity versus surfacing and evolving knowledge from the margins	• Encouraging expression versus risk of discomfort • Developing group agendas and ownership versus external agendas • Representing existing/superficial perceptions or evolving deeper insight • Enabling participant ownership of content during video co-construction • Participant's communicating in public space versus the risk of misunderstanding, entrenching difference, negative reactions, or backlash

5. GENERAL STRATEGIES FOR ETHICAL NAVIGATION OF PARTICIPATORY VIDEO PROCESSES

In this section, I illustrate some general strategies I have developed to ethically navigate participatory video processes, through projects that took place during the *Participate* research. *Participate* was initiated to bring the perspectives from some of the poorest communities into the UN deliberations that led to the SDGs. Researchers from the Institute of Development Studies (IDS), UK, collaborated with 18 partner organisations with long-term relationships with highly marginalised communities in 29 countries, and these partners undertook participatory research to explore the drivers and barriers to positive change in these contexts. During my role leading the *Participate* visual methods programme, I directly accompanied four partners in Kenya, India, and the Palestinian West Bank in using participatory video, both for community-led action research, and to open communication with influential policy-makers at national and global levels. Full details are beyond the scope of this chapter. To illustrate how participatory video processes can unfold, here, I draw predominately on one of the Kenyan fieldwork cases (Gathigi & Shaw, 2014). This involved a group called the Mathare Mappers – young people from an urban slum in Nairobi who were peer researchers on this project. I also draw on other *Participate* cases as comparison.

5.1. Supporting Autonomy Whilst Building Inclusion

Ethical practice in participatory video is fundamentally an ongoing relational matter, which plays out through progressive project interactions. The aim is to support participant's decision-making autonomy, which challenges the tacit assumption that they are powerless, and need protecting. The overall strategy I use is to progressively inform and support choices at each stage.

5.1.1. Progressive Processes of Informing and Supporting Participation Choices

The video-mediated action research during all *Participate* projects was conducted through local partners with long-term relationships with the marginalised communities. This was to ensure that the research was not extractive or superficial, and the groups could be supported afterwards. The first decision for potential participants in context is whether to join a project. The young women and men in the Mathare Mappers were already researching slum-dwelling issues through Global Positioning System (GPS) mapping with the local partner *Spatial Collective*, and were excited about using video in parallel. However, participatory video is a small-group activity with a maximum of eight people needed to ensure active involvement. In this case, we practitioners first facilitated a short participatory video process with a sub-group of Mappers. Then we accompanied this sub-group in running the same process with their peers to involve more young people.

As the Mathare Mappers were already doing GPS research, they were confident enough to try something new; but it is important to think carefully about how to engage less-confident people to avoid perpetuating marginalisation. At the same time, people's choices about whether to participate must be respected. Engaging people can happen through local gatekeepers, or progressively as a core group involves others, as they did in this case. Alternatively, a taster session can inform a decision to join a project. Sometimes, this happens informally through setting up a camera in a public space (e.g. the street or park), and offering people a turn, before inviting them inside to watch playback. On other occasions, more thought is needed: for example, a video project with the partner *Reality Check* in Indonesia started in the home of one of the poorest villagers, and worked out from there (Shaw, 2017a).

Participation choices are then progressive: next is choosing whether to participate in a project in order to learn how to use video, or explore issues in confidence. Consent to use the insight generated for research learning is then re-visited once participants have greater awareness of what is involved and more confidence to refuse. Consent to show video materials is approached when the externally focussed phase is reached, and discussions about what to communicate are informed by the experience of video recording and playback in the earlier phases (see further below). As consent is reviewed as the process unfolds, my strategy when submitting an ethics proposal is to include an ethics management plan. As well as anticipating risks, and detailing how these will be mitigated, this identifies key process stages or triggers when ethics and consent will be re-visited; and includes a schedule for regular ethics reviews, as well as a response plan in case ethical dilemmas arise that need attention. Success in implementing these plans lies in the interactional context established by the practitioner, which makes them active agents in the processes.

5.1.2. Practitioners Active Role in Generating an Enabling Context and Inclusive Dynamics

Extended participatory video processes in contexts of marginalisation aim for good, through progressively building people's sense-of 'can-do', understanding and capacity to influence. The intention is to generate inclusive group dynamics, which means everyone in the group has equivalent opportunities to express their ideas, be heard, and use the equipment. However, this aim confronts the group dynamic tendency for one or two people to dominate and other's to be left out (Shaw, 2016, p. 5). Navigating this effectively, to shift unequal power relations amongst participants, in order to address rather perpetuate marginalisation, necessitates a practice balance between supporting individual participant's choice, and intervening to establish and maintain inclusive dynamics. For example, handing out video equipment for people to use un-supported is likely to be a disempowering experience for many (Shaw, 2012b, p. 168). It maintains marginalisation by further increasing the capacities of the most confident, which is not fair to the least. Practitioners thus have an active role in maintaining safe space and inclusive dynamics. It is their support that determines whether the challenge of using or appearing on video will be a positive or negative experience for unconfident participants (Shaw, 2016). The details are beyond the scope of this chapter, but the Mathare group began using recording exercises like *statements in a round* (Shaw, 2017a, p. 20). This involves each participant speaking one-by-one as a microphone is passed round the group, and everyone takes turns using the camera each exercise. This stage is crucial to building the confidence of participants who are unused to speaking up, and none of this video material is shown outside the group following the basic principle outlined above.

Whilst turn-taking can back up the purpose of generating democratic space, it is practitioners who facilitate this – the technology doesn't do it on its own (Shaw, 2017a, p. 8). For instance, in Mathare young men repeatedly pushed forward to use the camera, and, as is usual, the practitioners needed to intervene firmly to ensure space for the women. As each individual has particular needs two practitioners are recommended. One can adapt equipment support one-to-one, whilst the other facilitates group activities. Experience also suggests that a more homogenous group is more inclusive, as the more heterogeneous a group, the harder it is to avoid some participants being disadvantaged during activities (Shaw, 2016).

Practitioner input is initially focussed on generating an inclusive context, in order to build individual and group agency and shared understanding about the issue or situation being explored. As a group develops a mutual agenda and their ideas about video stories or messages to make, the process is led by their interests. Nevertheless, it is still necessary for the practitioners to structure production activities given that the group members are inexperienced video-makers. This means there is always an element of co-construction to ensure successful video-making, but the balance of practitioner and participant direction is intended to shift as group capacities grow. This requires considerable reflexivity, given the potential for practitioners to drive content as well as structuring activities, and the external pressure to ensure there is an output to fulfil programme expectations (Shaw, 2012b, p. 52).

5.2. Anonymity versus Evolving and Surfacing Knowledge from the Margins

I have raised the risks of perpetuating marginalisation through neglecting grounded knowledge. I have also highlighted the inherent tension between supporting people to speak truth to power, and the risks of inappropriate exposure or negative consequences when using visual methods. I now specify some general strategies for navigating this ethically.

5.2.1. The Starting Point in Mitigating the Ethical Risks of Inappropriate Exposure

As specified in Section 4, a main strategy to mitigate the risk of video materials being shown before participants have enough experience to inform decisions is to clearly separate the initial group-building and exploration stages, from video's use to mediate external communication. During the initial stages, video is recorded solely to prompt refection and discussion in confidence. Later on a group may produce video material for specific external audiences. However, this requires that practitioners think carefully about where the research knowledge lies in the early stages (see Shaw, 2017c). Video recording followed by playback aids people in standing back from their experiences. This is important in deepening research knowledge, as participants can realise they didn't mean what they said initially, or can be prompted to think about their assumptions, social norms, missing perspectives, or why an issue is happening (see double-loop learning below). This means that key research insights can be in the prompted discussions, not the visual material recorded. This kind of learning can be included in research reports (assuming consent), without any need to show video materials. However, it is important to document these discussions as with other research methods. Additionally, funding agencies and other project stakeholders must be aware that no video materials will be shown unless or until the group chooses to, which might challenge their expectations because they may assume that there will be a video output from a video project. This was agreed with the *Participate* funders.

Another key aspect of my participatory video approach is that the group is the subject of their own videos (at least during early production). Every participant is part of the technical crew and appears in the recorded materials. This is different to typical video-making, where other people are the subjects of the producer's story. This strategy avoids the ethical risk of inexperienced participants recording others, without adequately informing consent processes. It also means if other people are videoed by the group later on, they are more aware of what it feels like to be recorded, and how to inform consent in a participatory manner.

5.2.2. Negotiating between Developing Group Purpose and the External Agendas

Both PAR and community-based participatory research are intended to follow community-led agendas and investigate topics that concern local participants (Burns, 2007; Minkler, 2004). The conundrum is that, in contexts of marginalisation, video projects are generally instigated through outside intervention by researchers or programme leaders from more privileged backgrounds. This provides a challenge to researchers wanting to investigate their own research questions with a group. One strategy is to develop broad enough research questions, within which groups can explore what matters most to them. The overall *Participate* research question asked how change happens and what prevents it, from the perspectives of people living in poverty. This left room for the Mathare group to focus on health and sanitation, and build insight relevant to the global research aim.

Some groups will already be working together, and have developed their own agendas. But, a newly formed group is unlikely to arrive with a pre-defined topic, and the project processes build shared purpose before video-making. Early participatory video exercises provide time for embryonic ideas to emerge, and help avoid dominant group members, the researcher, or external actors, driving the investigative direction. For example, the *shot-by-shot documentary* exercise (Shaw, 2017a, p. 21) assists in generating inclusive team dynamics through structured role rotation. All participants take a turn as director, camera operator, sound recordist, presenter, and floor manager, in order to produce one shot. They then swap round. Like a game of consequences, each shot is recorded one-by-one, which generates a documentary-style sequence in about an hour without pre-planning. This rapidly engenders a sense of accomplishment, as well as informing participant's awareness of what video production involves before their future planning. It also starts building their public performance capacities. Sufficient time to play creatively with video using similar exercises (e.g. shot-by-shot drama) reduces performance anxiety as people have fun improvising together. In addition, the group-building phase provides time for participant's agendas to emerge. In Mathare, the group had intended to explore health and sanitation only, but personal security came up repeatedly in the video exercises, so they added it as a major research focus.

5.2.3. Representing Existing Perceptions or Evolving Deeper Critical Insight

Another rationale for a period of structured exercises is to develop participant's perspectives on an issue through video-mediated exploration and reflective discussion. Few arrive with pre-formed opinions, and what they express initially can be relatively superficial or reflect damaging social norms or constraining assumptions. Both practitioners and participants have blind spots arising from their positioning, which takes time to unpack. In the participatory video context, to begin with participants can lack confidence to say what they really think, and group dynamics can limit open exchange (Cooke, 2001). This impacts research insights and can *do harm* through re-enforcing negative stereotypes and attitudes. This reflects the curtailment of deeper contextual understanding, which is often found in single-loop rather than double-loop[3] action research processes (Maurer & Githens, 2009, p. 278). Similarly, participatory video created through one iteration of production and reflection activities is more likely to lead to superficial content rather than richer insight.

For this reason, the audience for the first videos made is often the group themselves or their peers. At this stage in *Participate*, the Mappers spent nine months on horizontal-level (peers and wider community) group inquiry about health and sanitation and personal security (Shaw, 2017a, p. 26). They used progressive videoing and playback to mediate their explorations, and all sorts of video material resulted, including personal statements, contextual visuals, peer and community interviews, in-camera edited documentaries and dramas, and focussed messages.

The next stage involves critical reflection prompted by practitioners. Participant's might consider what was communicated and why, and whose views are missing. This can be combined with group sense-making to synthesise key research themes from the video materials, or a power analysis of participant's stories. In Mathare, on the second fieldwork visit we used a narrative arc to shift discussion from the issues identified to what participants thought could be done. Following this, the group was better prepared to make a video to stimulate discussion with an external audience, and we worked with them to make a video message to go to UN policy-makers.

5.2.3. Enabling Participant Content Authorship during Co-construction Processes

At least in the short and mid-term, marginalised groups are likely to need support to produce videos that communicate clear messages to external audiences. It is unrealistic to expect people with no prior experience to develop the same skills in less time as, for example, students on three-year media production courses. However, the editing process is a sticking point of involvement (Shaw, 2012a, p. 228), due to participation costs such as time/income loss. To address this, an 'in-camera editing' (Shaw, 2017b, p. 9) or 'no editing required' (N-E-R) (Mitchell, 2011, p. 74) approach is recommended. After storyboarding, video shots are recorded in order using the camera. Even if in-camera edited material is tidied up, or made more interesting with additional visuals, by someone else using computer editing software later, this can maintain group control because content choices are made by the group before recording. This strategy also aids sustainability as a time-efficient production process, which participants can use afterwards to make short videos on their mobile phones or cameras of adequate quality for local audiences.

As planning a complete video can be overwhelming, another strategy to enable participant authorship is to divide production decisions and tasks into manageable chunks. Videos can grow organically through iterative recording cycles, and group reflection on what else is needed (Shaw, 2012b, pp. 220–229). Practitioners also support participants in constructing compelling narratives and visual sequences, which is part of their input to bring unheard views into the public domain. In this way, I with *Participate* partner Real Time supported the Mathare Mappers to storyboard, record, and edit a video[4] for global policy-makers, which was incorporated into the *Participate* documentary, and the *Participate* online and New York exhibitions before the 2013 UN summit.[5] Their key policy message was:

> We cannot tackle wider social problems alone, so we want you to work with us not only on prioritising issues, but on implementing specific actions. (Mathare mapper in Shaw, 2017a, p. 29)

This is only a snapshot of the complex range activities with the Mathare Mappers in Kenya, but it is obvious that mediating iterative research and communication processes through participatory video is different to short-term group film-making as data generation. Instead these processes can potentially, as they unfold, establish relationships that in themselves shift marginalisation.

5.2.4. Navigating Audience Risks through Video Playback in Progressively Diversifying Spaces

Audience responses to videos made by marginalised groups are a relatively recent area of practice research. Showing videos to external audience(s) has the potential to communicate participant's messages and stories in succinct or compelling ways, increase audience understanding, and reposition marginalised groups more in influentially in external interactions due to presenter-audience role conventions (Shaw, 2017b, p. 14). Video, like other visual methodologies such as digital storytelling can generate empathy from an audience through human-to-human connection. The assumption is that leaders' attitudes are more likely to shift if they are personally moved (Jupp, Nusseibeh, Shahrokh, & Wheeler, 2014, p. 60), and stirring feelings rouse action that reading a report might not. However, there is no guarantee that increased empathy leads to greater responsiveness, and practitioners have also shown the risks of audience misinterpretation, provoking a negative reaction, or entrenching conflicting views (Kindon, Hume-Cook, & Woods, 2012; Shaw, 2012b, p. 252).

As mentioned above, building informed understanding of the potential consequences of being seen and heard on video is difficult because choices have to happen in advance of a screening event. In Section 4, I outlined how *audiencing,* as well as *production,* can be approached progressively. The strategy is for participants to develop their presentational skills firstly by showing videos to their peers and project supporters, then the wider community and then outwards to different external audiences. This can begin, for instance, with NGO allies and service providers, and then involve county, national, and global level decision-makers. This approach does not remove risk, but it better informs choices because consent can be iteratively appraised through participant's experience in different forums. This staging also opens a route for participants to develop their influencing capacities as they become social actors. In Mathare, the peer researchers first organised local screenings of the sanitation issues they had documented to stimulate discussion on what the community could do themselves. After this they went on to show their security video in wider community forums. Community members talked about how torches are used to turn security lights, and the young men decided to form patrol teams to create a safe corridor through the slum (Shaw, 2017c, p. 54). Later they ran an event to discuss solutions with local security stakeholders including youth and women's leaders, community elders, the local chief, and police and security agency representatives, who were keen to develop partnerships as the new Kenyan constitution requires that intervention is community drive. A participant reported:

> [...] the highest point is that we've been able to engage the community through ... coming together and doing something instead of waiting for the government to come and do so. (Mathare mapper in Shaw, 2017a, p. 26)

This process shifted local dynamics through nurturing these new social actors: both the core video group as they acted to motivate others, and those who were drawn in through video-mediated discussions. On an evaluation visit three years on, I found the same people still involved in local improvement projects. However, as expressed in the Mathare policy-focussed video, it is unrealistic to expect local action to solve structural issues without ongoing external support. Neither can audience reactions to participant's videos be predicted, as I discuss next.

5.2.5. Careful Thought on Communication Purpose and Framing for the Specific Audience

The Mathare mappers' engagement with local leaders was a success in fostering allies. However, a parallel *Participate* project with partner SEED in Nairobi, which explored the issues faced by children with disabilities and their parents (Shaw, 2015, p. 10), had two contradictory experiences with county-level decision-makers. In one screening, a decision-maker reacted angrily after watching the group's video, which finished with the statement that 'disabled children have been forgotten and ignored'[6] (Shaw, 2017a, p. 34). He felt the video presented the wrong impression of government efforts. In contrast, on another occasion the group used a video showing how their embryonic therapy centre had transformed life for a boy with cerebral palsy. This prompted an emotional response from leaders, and they were inspired to be further involved. These experiences emphasised the need to be clearer about the purpose of video-mediated dialogue, and whether it was to challenge leaders, to build understanding, to prompt discussion, or to foster allies (Shaw, 2017a, p. 41).

As the risks of audience misunderstanding or negative responses increase, the further removed screenings are from the context of production, it is also ethically important to consider how participant's videos are presented and framed. In terms of research insight, the different readings of visual materials and the discussions between audience members are key sites of knowledge generation (Rose, 2007), and the producer's interpretations are the closest to what was intended. A useful strategy to decrease the risk of misunderstanding is to not only contextualise videos through explaining the situation, the project aims and the way materials were generated, but also to include participant's views on what their video materials mean. This is whether videos are shown in live events or disseminated on online platforms. It is also important to consider the advantages and disadvantages of participants being present at the screening events, which will depend on context. Most *Participate* partners could not attend the UN exhibition, but there was an opportunity for attendees to interact with them in Skype sessions. In other cases, participants were not involved. For example, the disabled children in Kenya met leaders when they visited the locality for a video-mediated engagement event (Shaw, 2015, 2017, p. 31). However, they didn't attend the county-level engagements, as they were considered too physically demanding given impairment needs, as well as more interactively challenging. Instead, the adults brought them into the room via the videos.

5.2.6. Consent Processes Specify Different Usages, Audiences, and Dissemination Platforms for Video Materials

Even if approached progressively (as above), consent processes on participatory video projects need to be clear about the different usages of video materials, potential audiences, and distribution platforms. I suggest developing consent forms that specify all the possible uses of video material from transcription and analysis only, or the use of anonymised quotes in reports, to the variety of possible audiences. This might include screening at closed or open subject or methods training events, and to local, county, national, or global policy or academic audiences. It is also necessary to specify possible dissemination platforms including project websites or video sharing sites such as Vimeo, and who will have access. For example, if a video is uploaded to Vimeo it is possible to select whether the video will be visible to anyone open access, or only available to watch with a password or email link.

Whilst consent forms can include all these details, it would be unusual to discuss consent for all uses at one time. More usually, the group will re-visit consent at each progressive audience stage. As well as the experience of each screening informing the next decision, it may also be possible to show participants similar online platforms, or take them to training or academic events to inform understanding. It is also important to tailor consent processes to the people involved. Written consent should consider cognitive abilities and translate language or use pictures where appropriate, and written consent may increase misunderstanding or mistrust. In this case documented verbal consent processes would be better. Participants usually chose whether they want to be credited by name (which most do), but it is obvious that this needs to be a group decision. As there can be group pressure, the practitioner has a role in assisting the dynamics to ensure everyone is happy, as well as being reflexive about their own power to influence decisions, especially given external expectations or project commitments.

There are clearly limits to withdrawal of material from an edited video once it is completed and distributed. It is important that participants are aware that online material might be difficult to remove in the future, and they consider how they might feel about their communications in later years. Data protection requires that video materials are secured, and access controlled appropriately. In the context of General Data Protection Regulation (GDPR), it is proposed that robust data protection systems, and storage and access processes are set up. Raw video footage that could be re-edited or early process videos are not made public or deposited on open access platforms with only final videos with group consent stored in accessible research data banks. It is also suggested that a group's output videos are licensed using creative commons licensees that do not allow for re-editing.

To summarise, in this section I have covered some of the general approaches and strategies I developed to mitigate and navigate the risks of using participatory video with marginalised groups. However, this is not a practice blueprint as every group and situation is different, and there are still tensions to navigate in any participatory video process, which I now explore.

6. ADAPTING ETHICAL STRATEGIES TO THE SPECIFIC GROUP AND PARTICULAR CONTEXT

Marginalised groups are in reality not a unified category, and the term encompasses a very diverse range of people and situations. The strategies suggested for mitigating the ethical tensions pre-project, such as a group being the subject of their own video, provide a pragmatic beginning. But, what actually transpires as a project evolves will be dependent on four key factors, which delineate the territory in which practice decisions are made (see Table 3).

Tacking marginalisation usually requires action by external allies. It is necessary to support the least powerful groups in developing capacities and greater influence over their lives. This is the rationale for approaching participatory video as a facilitated process. However, the time needed and the number of process

Table 3. Factors Affecting How Participatory Video Can Evolve Ethically in Context.

More Time and Process Iterations Needed	Continuums of the Participatory Video Practice Territory	Less Time and Process Iterations Needed
Most hidden or stigmatised Groups	⟷ Nature of group	Group face discrimination, but have public voice
Fragile states, conflict contexts, and highly unaccountable governance	⟷ Particular context	Open and democratic public space – stable economy and politics
No previous involvement or capacity-building chances	⟷ Starting point of group members	Politically active – activists, advocate, or community leaders
Most painful, intimate, taboo, or contentious	⟷ Sensitivity of issue	Least personal or contentious

iterations depends on the nature of the group, the particular context, the starting point of the group, and the sensitivity of the issue, and this determines what is ethically appropriate at which stage.

6.1. Nature of Group and Starting Point of Group Members

The iterative videoing processes exemplified in this chapter may seem lengthy. However, the time needed and the progress possible depends on the group. For example, one *Participate* project with transgender communities in Chennai spent two days going through the confidential group-building phase, and then produced a video for UN policy-makers in four days. Transgender people in India are discriminated against, and often live in poverty; but it was possible to proceed rapidly because the participants were already politically active, and several ran advocacy and education organisations. It was similar with Bunge La Mwananchi in Nairobi (Shaw, 2017b): this ground-level social movement engages poor and marginalised citizens in political dialogue in informal spaces such as parks, markets and bus stops. The group had already carried out action research to explore what helped and hindered their success, and made a video to communicate the insights during a one-week participatory video process. They were able to achieve this partly because of the structured video production methods, but also because they were strong communicators already experienced in public expression (Shaw, 2017b, p. 9). Comparatively, during my time as co-director of Real Time, I supported a participatory video project with adults with learning difficulties for more than 15 years. During this extended period, the group made many videos, were involved in training other people with learning difficulties to use video, were assisted in running their own video organisation. However, they continued to need practitioner support (Shaw, 2012b, p. 262), and it would have set them up to fail to presume otherwise.

6.2. Challenging Contexts and Sensitivity of Issue

At the other end of the continuums in Table 3 are the most marginalised or stigmatised groups, the highly unaccountable or conflicted contexts, and the more sensitive issues. It doesn't mean that participatory video cannot be useful in building people's agency or sense of power-within (Shaw, 2017a, p. 10) on a short project. It just means showing materials may be inappropriate. For example, on the recent phase of *Participate*, the Egyptian partner collaborated with adults living with HIV/AIDs who are a highly hidden minority: many of the participant's families didn't even know of their health status due to stigma. Some participants were involved in participatory video and other visual methods, but it was decided too risky to show the material publically afterwards (Howard et al., 2018, p. 22). Nevertheless, research insight was still generated about the differences between group members due to infection pathways and gender, which meant some were more disadvantaged or hidden than others both within the communities and in terms of public health provision. This insight is valuable without any need to show the visual materials. There are also a range of other strategies for communicating findings generated through participatory visual processes. For instance, a participatory video project generated knowledge about life for children living with alcoholic parents, but the final video output used actors directed by the participants to communicate their narratives. Digital storytelling can also be used to tell a group or individual narrative, with pictures or photos to illustrate it, which is less revealing.

It is clear that the more painful, intimate, taboo, or contentious an issue the more time and more iterations of the process would be needed when using participatory video or digital storytelling to be ethically appropriate. For example, in a long-term research and communication project with people living with extreme violence and police corruption in South Africa, participants first made individual digital stories, then produced collective videos (Black, Derakhshani, Liedeman, & Wheeler, 2016), and later took part in deeper exploration to re-frame their experiences using hand mapping and further video production. Participants thought the strong group bond enabled them to surface painful experiences and transform their relationship to them, but recommended counselling back up for projects like this (Howard et al., 2018, p. 24). This project also emphasised the need think carefully about the purpose of video-mediated communication in highly conflicted contexts. The police reacted badly to the group video 'Gangsters in Uniform', because they were not pre-warned as to the purpose. This got in the way of building relations with potentially important allies and put group members at risk (Howard et al., 2018, p. 30). Neither is it realistic to think that marginalising attitudes, behaviours, and entrenched social norms will be overturned with one project. For example, de-notified tribes in India are extremely stigmatised as 'criminalised' by birth. Participatory research with them aimed to increase political leverage through aligning the project with a national campaign. However, practitioners reported that despite the strength of participant's digital stories, audiences interpreted people's predicament as their fault due to prejudice (Howard et al., 2018, p. 28). The purpose of these case examples in this section is to show that good intentions do not negate the inherent risks in attempting to not only do no harm, but to do good in these difficult contexts. At the same time, surfacing the difficulties in shifting the balance of social influence was part of the research learning.

7. LESSONS FOR NAVIGATING BETWEEN THE RISK OF NEGATIVE CONSEQUENCES AND BRINGING MARGINALISED PERCEPTIONS TO PUBLIC ATTENTION

In this chapter, I have addressed the knowledge gap on what situated ethics actually look like when using visual methodologies with marginalised groups. As the specifics are by definition contextual, I have tackled this through the exemplar of extended participatory video processes. However, there are insights that can be adapted to other visual methods. To better prepare researcher-practitioners for the reality rather than the ideals of participatory visual processes, it is important to be clear about the risks and challenges, and what helps to mitigating them. I therefore first defined two key participatory video practical balances, and I discussed some strategies I have developed to navigate them ethically as summarised in Table 4.

Table 4. Strategies for Navigating Participatory Video Processes Ethically.

Main Ethical Tension	General Strategies
Supporting decision-making autonomy whilst building inclusive dynamics	*Progressive process of informing and supporting choices* • Collaboration embedded through local partnerships, but time needed to reach the most marginalised • Taster sessions can inform choices about taking part in a project • Experience informs participation choices and consent throughout • Ethics management, monitoring, and evaluation plan can ensure regular ethics review and adaption *Practitioners active role to generate enabling context and dynamics* • Input to ensure safe space and generate supportive relational context • Use of video turn-taking processes to generate democratic space • Two practitioners enable better attention to diverse capabilities and needs • Homogenous group makes it easier to generate a more inclusive context, especially for the least confident groups at the early stages
Anonymity versus surfacing and evolving knowledge from the margins	*Initial mitigation of risks of inappropriate exposure* • Clear separation of video recorded to prompt refection in *safe* confidential space, from video's use to mediate external communication • Key insights from discussions and about dynamics need documenting, as they often contain the deepest insight rather than visual outputs • Needs for funder/project leads to agree that no video materials are shown unless/until the group decides to communicate externally • Group members are subjects of their own videos *Evolving new knowledge versus external influence* • Avoid imposing topics through broad research questions • Allow time for agenda emergence through creative video play • Structured videoing exercises enable ideas to develop and deepen through deliberation • In-camera editing and structured narrative construction process assist production success, and group authorship • Videos grow organically through iteratively production tasks • Deeper critical insight needs double-loop production processes *Navigating audience risks* • Showing videos to mediate dialogue in progressively diversifying spaces can build participant's communication capacities and inform their decisions on what is appropriate to show/say to different audiences • Need for clarity of communication purpose, and framing video materials using participant interpretations and prompting questions • Consider whether participants' will be involved in showing their own videos through attendance or not • Consent forms specify the different usages, audiences, and platforms for visual materials, and are adapted (Language and visuals) to group • Implications of consent forms are discussed in detail, and documented verbal consent processes when it is more appropriate • Need to navigate group decisions about consent for group productions, and be reflexive about group pressure or researcher/stakeholder coercion • Set up robust data protection systems, and only final videos with all group members consent are stored in accessible data banks

I would argue that the ethical tensions I have raised in this chapter are present in all participatory research with marginalised groups, not just when using visual methodologies. However, my research on participatory video identified that video intensifies both the process benefits and the risks (Shaw, 2017a, p. 40). This explains the enthusiasm of participatory researchers for visual methodologies, but also emphasises the need for caution. In a sense using visual approaches make 'visible' tensions that are intrinsic to social change endeavours in contexts of marginalisation, and somewhat counterintuitively this makes them easier to perceive and prepare for. As situated ethical practice is so context dependent, I finish with some questions for practitioners to consider when preparing to use participatory video, and by extension other visual methodologies, in similar circumstances.

- How will you progressively inform and support participant's choices about participation and public communication?
- How will you ensure clear separation of the use of visual methods to prompt research learning in safe confidential space, from their use in external communication?
- Given the particular group context and research issues, is there sufficient time for capacity-building and agenda development; and enough process iterations for critical insight and progressive audience decisions. If not, how will adapt the process to address the increased ethical risks?
- How will you ensure you (or the researcher-practitioners you involve), are able to support participant's production and narrative skills, and, most importantly, have the relational skills and approach necessary for process facilitation?
- Have you considered the purpose of using visual methods at each process stage? How will this inform your process and session workshop plans? For example, in using participatory video, will group members all take turns using the equipment, and appearing as subjects of their own videos?
- How will you navigate the balance between process and product needs given the purpose?
- How will you document the research knowledge that arises from discussions mediated by visual creation, or is revealed in the project dynamics?
- What are the considerations in showing and disseminating videos or other visual outputs made by marginalised groups? How will you approach anonymity and consent, and the framing of materials for different audiences?
- What will your process be for ongoing review of ethics practice, re-visiting consent, and responding to ethical dilemmas as they arise?

NOTES

1. Researchers using participatory methods, including participatory visual methods, facilitate participants in participatory research activities, as practitioners of the methods. In this chapter, following usual conventions, I refer to them as researcher-practitioners or simply practitioners.
2. See more at https://participatesdgs.org/

3. Single loop learning involves adapting practice according to the outcome of actions, whereas double-loop learning involves reflecting on the underlying assumptions or root causes of problems (Argyris, 2005).
4. You can see *Work With Us* at https://vimeo.com/showcase/4488354/video/80075380.
5. You can see *Working together for Change* at https://vimeo.com/210789695.
6. You can see the video at https://vimeo.com/210791930.

REFERENCES

Argyris, C. (2005). Double-loop learning in organizations: A theory of action perspective. In K. G. Smith & M. A. Hitt (Eds.), *Great minds in management: The process of theory development* (pp. 261–279). Oxford, New York, NY: OUP.

Banks, M. (2001). *Visual methods in social research*. London: Sage.

Beauchamp, T., & Childress, J. (2001). *Principles of biomedical ethics*. Oxford: Oxford University Press.

Black, G., Derakhshani, N., Liedeman, R., & Wheeler, J. (2016). *What we live with everyday is not right: Partnerships for accountability and safer cities in South Africa*. Cape Town: Sustainable Livelihoods Foundation.

Boxall, K., & Ralph, S. (2009). Research ethics and the use of visual images in research with people with intellectual disability *Journal of Intellectual & Developmental Disability*, 34(1), 45–54

Burns, D. (2007). Systemic Action Research: A Strategy for Whole System Change. Bristol: The Policy Press.

Burns, D., Howard, J., Lopez-Franco, E., Sharokh, T., & Wheeler, J. (2013). *Work with us: How people and organisations can catalyse sustainable change*. Brighton: IDS.

Burns, D., Ikita, P., Lopez-Franco, E., & Shahrokh, T. (2015). *Citizen participation and accountability for sustainable development*. Brighton: IDS.

Capstick, A. (2012). Participatory video and situated ethics: avoiding disablism. In E.-J. Milne, C. Mitchell, & N. de Lange (Eds.), *The handbook of participatory video* (pp. 269–282). Lanham, MD: Altamira Press.

Clark, A., Prosser, J., & Wiles, R. (2010). Ethical issues in image-based research. *Arts & Health, 2*(1), 81–93. http://dx.doi.org/10.1080/17533010903495298

Cooke, B. (2001). The social psychological limits of participation?. In B. Cooke & U. Kothari (Eds.), *Participation: The new tyranny?* (pp. 102–121). London: Zed Books.

Flicker, S., Travers, R., Guta, A., McDonald, S., & Meagher, A. (2007). Ethical dilemmas in community-based participatory research: Recommendations for institutional review boards. *Journal of Urban Health, 84*(4), 478–493.

Flyvbjerg, B. (2001). *Making social science matter: Why social inquiry fails and how it can succeed again*. Cambridge: Cambridge University Press.

Gathigi, M. K., & Shaw, J. (2014). Using participatory video in Nairobi, Kenya to enable community-led change—Methods and challenges in using participatory processes in diverse contexts: Bridging the gap between community reality and policymaking. T. Shahrokh & J. Wheeler (Eds.), *Knowledge from the margins: An anthology from a global network on participatory practice and policy influence* (pp. 27–31). Brighton: IDS.

Gilligan, C. (2008). Moral orientation and moral development. In B. Alison & C. J. Cuomo (Eds.), *The feminist philosophy reader* (pp. 467–478). Boston: McGraw-Hill.

Gubrium, A. C., Hill, A. L., & Flicker, S. (2014). A situated practice of ethics for participatory visual and digital methods in public health research and practice: A focus on digital storytelling. *American Journal of Public Health,104*(9), 1606–1614. doi:10.2105/AJPH.2013.301310

Gubrium, E., Harper, K., & Otanez, M. (2015). *Participatory visual and digital research in action*. Walnut Creek, CA: Left Coast Press.

Howard, J., López Franco, E., & Shaw, J. (2018). *Navigating the pathways from exclusion to accountability: From understanding intersecting inequalities to building accountable relationships*. Brighton: IDS.

Jupp, D., Nusseibeh, L., Shahrokh, T., & Wheeler, J. (2014). How change happens: Pulling together and closing the gap. In T. Shahrokh & J. Wheeler (Eds.), *Knowledge from the margins: An anthology from a global network on participatory practice and policy influence* (pp. 59–63). Brighton: IDS.

Kabeer, N. (2010). *Can the MDGs provide a pathway to social justice? The challenge of intersecting inequalities*. New York, NY: United Nations Development Programme.

Kabeer, N. (2016). "Leaving no one behind": The challenge of intersecting inequalities. In ISSC and IDS (Eds.), World social science report 2016: Challenging *inequalities: Pathways to a just world*, pp. 55–58. Paris: UNESCO Publishing.

Kemmis, S., & McTaggart, R. (2005). Participatory action research: Communicative action and the public sphere. In N. K. Denzin & Y. S. Lincoln (Eds.), *The sage handbook of qualitative research* (3rd ed., pp. 559–604). Thousand Oaks, CA: Sage Publications.

Kindon, S., Hume-Cook, G., & Woods, K. (2012). Troubling the politics of reception in participatory video discourse. In E.-J. Milne, C. Mitchell, & N. de Lange (Eds.), *The handbook of participatory video* (pp. 349–364). Lanham, MD: AltaMira Press.

Law, J. (2004). *After method: Mess in social science research*. London: Routledge.

Lewin, T. (2011). Digital storytelling. *Participatory Learning and Action, 63*, 54–62.

Manzo, L., & Brightwell, N. (2007). Toward a participatory ethics. In S. Kindon, R. Pain, & M. Kesby (Eds.), *Participatory action research approaches and methods: Connecting people, participation and place* (pp. 33–40). New York, NY: Routledge.

Maurer, M., & Githens, R. P. (2009). Toward a reframing of action research for human resource and organisation: Moving beyond problem solving toward dialogue. *Action Research, 8*(3), 267–292.

Minkler, M. (2004). Ethical challenges for the "outside" researcher in community-based participatory research. *Health Educational Behavior,31*, 684. doi:10.1177/1090198104269566

Mistry, J., Berardi, A., Tschirhart, C., Bignante, E., Haynes, L., Benjamin, R., ... De Ville, G. (2016). Community owned solutions: Identifying local best practices for social-ecological sustainability. *Ecology and Society, 21*(2), 42. http://dx.doi.org/10.5751/ES-08496-210242

Mitchell, C. (2011). *Doing visual research*. London: Sage publications Ltd.

Mitchell, C., De Lange, N., & Moletsane, R. (2017). Participatory visual methodologies: Social change, *community and policy*. London: Sage publications Ltd.

Papademas, D. (2004). Editor's introduction: Ethics in visual research. *Visual Studies, 19*(2), 122–126.

Pink, S. (2007). *Doing visual ethnography* (2nd ed.). London: Sage.

Prosser, J. (2000). The moral maze of image ethics. In H. Simons & R. Usher (Eds.), *Situated ethics in education research* (pp. 116–132). London: Routledge.

Prosser, J., & Loxley, A. (2008). *Introducing visual methods*. National Centre for Research Methods Review Paper No. 010.

Rose, G. (2007). *Visual methodologies* (2nd ed.). London: Sage.

Shahrokh, T., & Wheeler, J. (Eds.). (2014). *Knowledge from the margins: An anthology from a global network on participatory practice and policy influence*. Brighton: IDS.

Shaw, J. (1986). Process work and community video practice. *Independent Media, 57*, 7–8.

Smith, L. (2008). How ethical is ethical research? Recruiting marginalised, vulnerable groups into health services research. *Journal of Advanced Nursing, 62*(2), 248–257.

Shaw, J. (2012a). Beyond empowerment inspiration: Interrogating the gap between the ideals and practice reality of participatory video. In E. J. Milne, C. Mitchell, & N. de Lange (Eds.), *Handbook of participatory video* (pp. 225–241). Lanham, MD: Altamira Press.

Shaw, J. (2012b). Contextualising empowerment practice: Negotiating the path to becoming using participatory video processes. Ph.D. thesis, Institute of Social Psychology, LSE. Retrieved from http://etheses.lse.ac.uk/400/

Shaw, J. (2015). Re-grounding participatory video within community emergence towards social accountability. *Community Development Journal, 50*(4), 624–643. doi:10.1093/cdj/bsv031

Shaw, J. (2016). Emergent ethics in participatory video: Negotiating the inherent tensions as group processes evolve. Area, *48*(4), 419–426. doi:10.1111/area.12167

Shaw, J. (2017a). *Pathways to accountability from the margins: Reflections on participatory video practice*. Making All Voices Count Research Report. Brighton: IDS. Retrieved from https://opendocs.ids.ac.uk/opendocs/bitstream/handle/123456789/13149/PartipVideos_Report_Online.pdf

Shaw, J. (2017b). *Connecting communication: Using video to open spaces and mediate exchange between Kenyan grass-roots activists*. Making All Voices Count Research Report, Brighton: IDS. Retrieved from https://opendocs.ids.ac.uk/opendocs/bitstream/handle/123456789/13339/MAVC_RR_Shaw_BLM_final.pdf

Shaw, J. (2017c). Where does the knowledge lie in participatory visual processes. *Visual Methodologies*, *5*(1), 51–58 (4th international visual methods conference special issue). Retrieved from http://journals.sfu.ca/vm/index.php/vm/article/view/99/pdf

Shaw, J. (2021, forthcoming). Extended participatory video research processes. In D. Burns, J. Howard, & S. Ospina (Eds.), *Sage handbook of participatory research*. London: Sage.

Shaw, J. & Robertson, C. (1997). *Participatory Video: A practical guide to using video creatively in group development work*. London: Routledge.

Simons, H., & Usher, R. (Eds.). (2000). *Situated ethics in educational research. London: Routledge* Falmer.

Tracy, S. J. (2013). *Qualitative research methods: Collecting evidence, crafting analysis, and communicating impact*. London: Wiley-Blackwell.

Walker, R., Schratz, B., & Egg, P. (2008). Seeing beyond violence: Visual research applied to policy and practice. In P. Thomson (Ed.), *Doing visual research with children and young people* (pp. 164–174). Abingdon: Routledge.

Wiles, R., & Boddy, J. (2013). Introduction to the special issue: Research ethics in challenging contexts. *Methodological Innovations Online*, *8*(2), 1–5.

Wiles, R., Crow, G., Heath, S., & Charles, V. (2007a). The management of confidentiality and anonymity in social research. *International Journal of Social Research Methodology*, *9*(2), 83–95.

Wiles, R., Crow, G., Charles, V., & Heath, S. (2007b). Informed consent in the research process: Following rules or striking balances? *Sociological Research Online*, *12*(2), 1–18.

Wiles, R., Prosser, J., Bagnoli, A., Clark, A., Davies, K., Holland, S., & Renold, E. (2008). *Visual ethics: Ethical issues in visual research*. National Centre for Research Methods, Methods Review Paper 011.

PART III

CONTEXT AND REPRESENTATION

CHAPTER 7

ETHICS AND THE ARCHIVE: ACCESS, APPROPRIATION, EXHIBITION

Ciara Chambers

ABSTRACT

Archive footage is now a staple of much cinematic and broadcast production. This chapter explores some of the ways in which archival material has been recycled and considers some of the tensions between filmmakers, archivists, and audiences throughout the process of research, production, and screening. It considers some of the controversies associated with the repositioning of material in short-form, narrative and documentary filmmaking, particularly in relation to content that was never intended for exhibition in the public sphere. Drawing upon Benjaminian ideas of accessing authenticity in a form that has been reproduced, it considers the responsibility of both filmmaker and viewer in critiquing moving image content that has borrowed, self-consciously or surreptitiously, from earlier filmic forms. It concludes by making recommendations for an ethical approach to recycling archival material in research contexts that are pertinent to the burgeoning field of academic creative practice, with a particular focus on the stakeholders involved and a reasonable contextual positioning of the source material in its remediated form.

Keywords: Archives; appropriation; remediation; recycling; amateur film; creative practice

Ethics and Integrity in Visual Research Methods:
Advances in Research Ethics and Integrity, Volume 5, 133–151

ISSN: 2398-6018/doi:10.1108/S2398-601820200000005013

Imagine you are on holiday and on one of your long, languorous days you wander into an art gallery. Amidst photographs, paintings, and sculptures, a centrepiece emerges. It is an installation: a white wall on which flickering images fluctuate hypnotically. You watch the dreamlike bricolage with vague interest. Suddenly, your heart skips a beat. Was that …? No, there must be some mistake. Surely not. It is impossible. But wait, there it is again. You catch interrupted glimpses of a child on a beach. This child is familiar to you. You wait for another image. The child appears again, at a birthday party. Your breathing quickens. You know this person. You catch your breath as you see two more images of your youthful self-appear and fade before your eyes.

Perhaps in the age of digital proliferation of images of the self, this scenario does not seem too far-fetched. It is easy for anyone to take images of individuals, all too keen to share their experiences online, and use them as they please. As renowned filmmaker Alan Gilsenan (2012) observes:

> Never before have we been so observed. So recorded. So documented. On every street, in every parking lot, in every shopping centre. Online and elsewhere. In reality and in virtual reality. Each movement. Each translation. Each encounter. And never before have we documented *ourselves* so much. Recorded our every moment. On our phones, our cameras, our voice-recorders, our social media. We are constantly watching and listening. Ourselves and others. Each and every minute. And we are being watched and listened to. Nearly all of the time. Yet in most of these exchanges our consent is never clearly or consciously sought …. We are all slowly becoming the unknowing subjects in some strange abstract film, an unformed and shapeless documentary of our social netherworld. (p. 57)

Before the advent of the internet, the scenario described above would have been chilling for many individuals whose family albums and home movies were safely stored in the attic and never shown beyond the private sphere of the home, but the boundaries of privacy have now been redrawn, with the result that private images may now easily end up in the public domain.

For one anonymous woman, who may have visited the Centre national de l'audiovisuel (CNA) in Luxembourg in 2010, the imagined scenario above could have been all too real. The CNA hosted an exhibition between March and June that year called *I was here*. It presented 'different approaches on the subject of tourism, based on a photographic heritage that seems innocuous, anonymous, forgotten and often, from a technical point of view, of average quality' (Walerich in Kmec, & Thill, 2012, p. 193). Amidst the images taken from public archives, a section of the exhibition titled *in almost every picture #1* was devoted to found photographs. In this section, images with no known origin were selected and presented by artist Erik Kessels. The collection included images from 1956 to 1968 of the same woman, posing in different holiday locations. As curator Michèle Walerich (2012) describes:

> we observe her gradual ageing, witness the subtle changes in her person, from the variations in her hairstyle and taste in fashion, to how her husband's gaze upon her alters. (p. 194)

There is a haunting nostalgia in the images which present the evolution of a young girl into a more sedate woman and the viewer's imagination stretches to fill in the gaps. The photographs become

> an inexhaustible target for speculation, a fruitful terrain for inventing and reinventing their second lives. Their pureness of intention, their imperfection and flaws, have something salutary about them, when compared to the slick and over-processed visual world we are constantly exposed to. (Walerich, 2012, p. 194)

But it is exactly this slick and over-processed visual world that might forgive the exhibition of these images in the public sphere, given the obsessive nature of digital citizens who document every trip, meal and activity and share their movements freely with the world. This woman lived in an age when this was not only undesirable, but impossible. What would she think of this exhibition and does it matter? Do we have the right to look at her and recreate or reimagine her personality, relationships and existence for ourselves?

When teaching students about archival appropriation, the borrowing of images for film and television production, I show a clip from YouTube of a 1941 wedding video. The happy couple emerges into an Amsterdam street and a young girl watches them from above, leaning out of a second-floor window. After viewing the film, I ask the group for their thoughts on why humans mobilise the most recent technology available to them to capture these moments: weddings, birthdays, and holidays. Then I reveal the identity of the teenager watching the happy couple. She is Anne Frank. The discussion changes. Inevitably there is a morbid fascination in considering that we know what is about to happen to her, but she does not. We then talk about whether it was ethical to publish her private diary for the world to read and consider that perhaps because her father made this decision and not a stranger, that we can accept his choice. But are we invading her privacy? Reading her words and watching these images without her expressed permission? What about images of the emaciated bodies of the Holocaust? No one asked their permission to circulate images of their suffering, humiliation, and torture; would any one of us want to be viewed like this after death? Coming back to these images of Anne Frank, when it is learned that the Anne Frank House uploaded the footage to YouTube and that the organisation has a channel dedicated to the teenager's memory with a range of audiovisual material that is contextualised for viewers, there is a certain sense that this seems acceptable practice (Gabbatt, 2009). Context gives us a sense of satisfaction that no ethical rules have been broken. The provision of context is perhaps the most significant factor in appropriate repositioning of a cultural artefact. But approaching context, as we shall see, can be riddled with complications.

RECYCLING THE ARCHIVE: ACCURACY, AUTHENTICITY, AND ENTERTAINMENT

The archive is a minefield. Reusing it opens up a range of complex questions related to power, agency, and control. It is also a living and ever-expanding repository of personal and collective memories. Today's news will become library footage or 'archive' tomorrow. The archive has spilled out of protected institutions and into the public domain and 'the distinctions between archives, libraries, collections and other gatherings of objects, including virtual objects in digital archives, have increasingly blurred' (Baron, 2014, p. 2). Many public archives have digitised their collections (or at least their catalogues) and made them available online. This makes the job of a film or television researcher much easier than before. No longer do they have to book a viewing in, for example, the Imperial War Museum or the British Film Institute to search for images for inclusion in a documentary. No longer do they have to wait for an experienced archivist to lace up a Steenbeck for them. They can view and pay for the images at home and then order high resolution copies to be shared with them by file transfer. With copyright clearances in place, all the legalities have been taken care of and the footage is ready to use. But in many cases, the way in which footage is appropriated raises significant questions for archivists, filmmakers and audiences. As Jaimie Baron (2014) suggests:

> the ideas of the location, provenance and authority of an archive have become increasingly uncertain as online digital archives are constituted and accessed not only by institutions but also by individuals and groups all over the world. (p. 7)

The broader discourse of historical cinema often explores the tensions between accuracy and entertainment values, particularly when depicting real events or individuals. Many of these debates focus on the 'authenticity' of what is presented to audiences. With digital proliferation and the ease with which archive film footage may be recycled, there is a heightened focus on the evolution of a piece of film from its original production context to its remediated form. As Walter Benjamin (1935) notes 'technical reproduction can put the copy of the original into situations which would be out of reach for the original itself' (p. 733). This is particularly pertinent to the movement of the amateur film from the domestic sphere of the home to the public sphere of broadcast or cinematic exhibition. Benjamin also suggests that 'the presence of the original is the prerequisite to the concept of authenticity' which can be extrapolated to invoke a notion of contextualisation whereby the original footage is acknowledged by reference to the spirit in which it was made or originally exhibited. Imbuing the recycled form with some sense of its original context can invite the audience to critique how it has been reused, thus reflecting upon the material's journey through various formats and 'reactivat[ing] the object reproduced' (Benjamin, 1935, p. 734).

In a similar vein, Nick Couldry draws on Émile Durkheim's idea of 'civic morals' when formulating an ethical approach to media culture (Durkheim, 1992, cited in Couldry, Madianou, & Pinchevski, 2013, p. 39). Suggesting that accuracy, sincerity, and care are key values underpinning media ethics, he raises this debate for both professional producers and 'anyone who acts with and through media, including digital media platforms' (Couldry et al., 2013, p. 40). Couldry thus shares ethical responsibility between creator and audience, encoder and decoder. As mentioned elsewhere in the text:

> audiences are not merely spectators, but witnesses who find themselves having to negotiate between, and respond to, the moral issues that arise from the seemingly innocent act of everyday involvement with the media. (p. 12)

Within the academy, there is a heightened responsibility in both the production of media works to be consumed by audiences and in the education of students to reconsider their positioning as viewers. The academy then, is a space whereby best practice in media ethics can be developed as an intellectual pursuit and one which challenges the abuses and hierarchies of media production, educating both audiences and producers. In explaining the ethical responsibilities of the media producer, Couldry (2013) defines both accuracy and sincerity as 'truth-telling' virtues (p. 48):

> Accuracy is the disposition not so much to always hit the truth (truth is too complex an achievement for that to be a plausible characteristic of anyone) but to make the 'investigative investment' (Williams, 2002, p. 124) required for generally obtaining the truth. Sincerity, by contrast, is the disposition to make appropriate efforts to ensure, as far as possible, that what one says is consistent with what, more broadly, one believes.

Accuracy then, when it comes to the archive, could translate into an attempt to represent the original archive material *truthfully*; in other words there is no agenda deliberately to mislead the audience in relation to what they are viewing. This could easily be replaced by an attempt at authentically representing the material that is recycled: making its context and its appropriation clear and not attempting to mislead the audience into believing it is something it is not. Couldry's further call to `take care with respect to the effects of our media communications as they circulate' is due to

> the fact of our commonly experienced connectedness, the common fabric of a mediated world, which makes all of us vulnerable to each other. Misrepresentations (and the regular patterns or gaps in media representations) can always do harm. (p. 52)

Thus, both sincerity and care are encapsulated within the desire not to mislead and avoid any intentional harm to viewers or stakeholders.

With these values in mind, this chapter will now explore some case studies of archival appropriation by artist-practitioners with a view to understanding how academic creative practice might learn from professional production.

DIRECT REUSE

Archive film has formed the basis of news and documentary production from the origins of cinema. Before television, it was standard practice for cinema newsreels to borrow from footage held in their libraries to illustrate news reports and many television stations still operate on this principle. Productions are often marketed on the basis of screening 'never before seen' footage and documentary filmmaker Ken Burns became so known for his creative appropriation of the archive, that the term 'Ken Burns effect' came to be associated with panning and zooming across still images to add the illusion of animation, particularly when moving images of the topic depicted were not available.

While there has been a steady stream of historical productions using archive footage along with talking heads, it is interesting to note some recent documentaries that create heroic narratives more akin to dramatic reconstruction, by relying solely on archival material. Director Asif Kapadia has proven to be particularly adept at this task in his documentaries *Senna* (2010) and *Amy* (2015), so much so that it was this feature of Kapadia's work that was used for marketing purposes for his most recent offering *Diego Maradona* (2019) which, the general public was told, was based on 'more than 500 hours of never before seen archive footage' (Interview with the archive researchers). The differing responses to the first two films make for an interesting comparison in relation to the ethical recycling of the image of a deceased person.

Described as 'a discreet hagiography' by the *New York Times* (Holden, 2011), *Senna* introduces us to our hero, invites us to watch him grow and evolve through home movie footage and television broadcasts, offers us an antagonist in the form of Alain Prost and builds, like any good narrative, to an excruciating climax. And as with all conventional narratives, equilibrium is restored at the end of the film and the audience is satisfied with the closure offered: the dangerous rivalry with the anti-hero is overcome; we can celebrate the memory of this famous driver in the knowledge that racing is now a much safer sport and many of those depicted seem to have gained personal development through their interaction with the story's hero. It was Kapadia's unique approach that allowed him access to the footage that had been denied other directors, as he outlined in an interview in *The Guardian*:

> Lots of filmmakers over the years approached the Senna family, Oliver Stone, Michael Mann and I'm pretty sure Ridley Scott all approached and were told, 'no'. Antonio Banderas wanted to play Senna .... The main thing was they all wanted to make a film about his final weekend at Imola in 1994. The family didn't want that. They preferred what we wanted to do, which was a three-act drama celebrating his life, from archive footage. (Kapadia quoted in Jeffries, 2011)

Here it seems like the ethical approach of consulting and working with the wishes of the family of the person portrayed won the day. The film was a huge critical and popular success, winning two BAFTAs and a host of other awards. It was also the first of a trilogy of 'true fiction' produced by Kapadia relying on the traditional three-act structure involving the conscious employment of exposition, jeopardy, and resolution, that is a standard feature of narrative filmmaking.

Amy was compiled using a similar methodology to that employed for *Senna*. Apart from a few drone shots, it fuses a range of amateur and professional archive footage with audio interviews with friends and colleagues of singer Amy Winehouse. The depiction in this case is, like Winehouse's public image, far from discreet and there is a strong focus on the singer's drug and alcohol addiction and her public decline. As was the case with Senna, the Winehouse family was approached with several project ideas in relation to a film about the singer's life, but relatives considered many of them 'trashy'. Kapadia's more considered approach again won out, but this time the family, and in particular Winehouse's father Mitch, was unhappy with the result. After initially supporting the project, the family damned the film on completion, claiming 'it is both misleading and contains some basic mistruths'. Mitch Winehouse, particularly upset at his own depiction in the film, described his comments to the production team after a private screening:

> I told them that they were a disgrace. I said: 'You should be ashamed of yourselves. You had the opportunity to make a wonderful film and you've made this'. (Winehouse quoted in Saner, 2015)

The production team released a statement, suggesting the portrait that emerged in the film came from the interviews they conducted:

> When we were approached to make the film, we came on board with the full backing of the Winehouse family and we approached the project with total objectivity, as with *Senna*. During the production process, we conducted in the region of 100 interviews with people who knew Amy Winehouse; friends, family, former partners and members of the music industry who worked with her. The story that the film tells is a reflection of our findings from these interviews. (Kapadia quoted in Saner, 2015)

It is clear that at some point during production, Kapadia deviated from the wishes expressed by the Winehouse family. From an ethical standpoint, this example would demonstrate that respecting the wishes of the deceased person's family may be preferable in terms of garnering public approval of a production. Family consent seems to limit public backlash, although of course a family will often attempt to protect its own mythology and limit uncomfortable explorations of any wrongdoings by one of its members, thus causing a dilemma for the director of a biopic in documentary or narrative form. Catering to the desires of the family may impinge on the 'truth' that may be told about a subject, particularly if family members restrict or withhold information.

There is also another element in terms of the representation of those close to the main subject, particularly those still living. Winehouse was concerned that he was depicted as 'greedy and uncaring' and he is shown as just one of the men in the singer's life who were problematic. Even the setting up of Alan Prost in *Senna* as the traditional villain is not particularly negative as, after some useful and dramatic rivalry, he is ultimately redeemed. Interestingly, Kapadia discussed his approach to building up trust with Amy Winehouse's inner circle:

> That was really the big job for me, that was to earn the trust of these people who have never spoken or given an interview, were in a lot of pain, angry with the media, angry with everyone really, for what had happened. (Kapadia quoted in Amy Interview Special, 2015)

Clearly, there was a split in the strength of this relationship between friends and immediate family members, which may have been associated with the depiction of certain figures in the film. Family consent extends not only to the portrayal of the dead person, but to any reflection on the relationship of that person with those who are still alive. But the approval of the most important people to consider in relation to these representations has never been sought. What would Aryton Senna or Amy Winehouse think of these depictions? As for Diego Maradona, it has been reported that he has not yet viewed Kapadia's portrait of his life, but the effects in this film are largely a combination of the hagiographic treatment of Ayrton Senna with the exploration of the demise of Amy Winehouse. Thus Maradona is both a heroic and a tragic figure and depicted as a victim of his circumstances and the ever-changing whims of the tabloid media.

Ken Loach has garnered much critical acclaim for his oeuvre, grounded in social realism, and his reputation is further enhanced by his lifetime devotion to social activism. From an ethical perspective, he takes on the issues of the communities he portrays and clearly holds a strong personal belief in the transformative power of film. His provocative film about the Irish Civil War, *The wind that shakes the barley* (2006), was heavily criticised for its perceived Republican tendencies. In an unusual aesthetic choice for Loach, he appropriates newsreel footage in a scene set in a cinema. The audience watches a report of the signing of the notorious

Anglo-Irish Treaty. Heralded by the British as a solution to the 'Irish Question', the Treaty plunged the country into a bitter civil war. Remarkable footage covering the story exists in the form of a film released by both Topical Budget and Pathé: *Peace Council at the Palace* (8 December 1921). Loach combines some of the iconic shots from this film with segments from other newsreels released over several months to form his own hybrid news film to which he adds some newly written intertitles to provide narrative cohesion. The result is a compendium of information about the Irish situation, neatly conveyed through the manipulation of real archive footage.

Loach's appropriation of the footage is not strictly accurate and certainly leads viewers to believe they are watching an authentic news report from December 1921. But while Loach's temporal elisions might draw the scrutiny of archivists, the average viewer is not misinformed *per se*. The images are presented much in the spirit of the original exhibition conditions: a cinema with a packed audience, listening to the standard musical accompaniment of the silent era alongside the voices of literate patrons announcing intertitles for those who could not read and absorbing the loud responses of viewers to the images unfolding before them. However, audiences were seldom informed about events via the cinema in the early 1920s, and indeed the newsreels tended to provide moving images depicting events that viewers were already familiar with through newspapers and word of mouth. With an event as significant as the signing of the divisive Irish peace treaty, audiences would have most certainly had prior knowledge through the dissemination of information by the Irish delegates. Loach's historical bricolage offers an interesting aesthetic intervention into the narrative, accurately representing Irish dissatisfaction with the conditions of the Treaty and offering the extra-diegetic audience a perspective on the conflict in store for the viewers depicted within the frame (Chambers, 2018, pp. 265–266).

INDIRECT REUSE

Often a filmmaker will consult archival documents or footage for the purposes of contextual research rather than direct reuse. This leads to borrowings of a different status than that of direct appropriation. Peter Mullan, director of *The Magdalene sisters* (2002) consulted clerical material held in the Irish Film Archive while writing his script for the film. In particular, he viewed material from the Delaney collection, a series of films shot by Father Jack Delaney of his inner city Dublin parish in the 1930s. It shows tenement life, religious parades and scenes from a local convent which housed women known as 'Magdalenes'. The Magdalene laundries were workhouses run by the Catholic Church to incarcerate women who had become pregnant outside marriage or committed other alleged misdemeanours deemed to be sexually inappropriate. In some cases, a fear that a woman was capable of sexual misconduct was enough to condemn her to one of these institutions. The women were subjected to harsh working conditions and those who bore children had them removed from them soon after birth; many of the infants were adopted in Ireland and abroad and scant records were kept, making it difficult for birth parents and children to reunite later.

The widespread abuse committed against Magdalene women has been extensively explored in Irish mainstream media and one woman's struggle to find her son was depicted in Stephen Frears' film *Philomena* (2013). The Irish government has paid millions in compensation to hundreds of survivors under the Magdalene Laundries Restorative Justice Ex-Gratia Scheme established in 2013. Delaney's film of course shows no such abuse, and the Magdalenes are represented in a way that suggests they are well treated: they are uniformed but well dressed and shown smiling at the camera, playing games and conversing with ease with nuns and priests. Mullan's portrayal of a litany of horrors – sexual abuse, beatings, humiliation, and cruelty – contradicts Delaney's representation and yet the two films are linked in a reference point which raises significant questions about the role of archival research in historical cinema.

Sunniva O'Flynn, then Curator of the IFI Irish Film Archive, described her experience of watching Mullan's film to ascertain if the archival research would be visible in the final production.

> It was most strikingly evident in the character of Father Fitzroy (Daniel Costello), an amateur film enthusiast who in one scene records on his 16mm camera the women (nuns and Magdalenes) in the grounds of the convent in a direct re-enactment of the Father Delaney film. The scenes which follow this reconstruction depict Father Fitzroy sexually abusing a young girl. The priest is an amalgam, conflated from unrelated facts. The characters of the four women in the film are also inspired by but not directly based on real women whose lives within Magdalene institutions were the subject of the documentary *Sex in a Cold Climate*. There is room for some confusion as to the extent of the fictionalisation in the film, particularly with regard to the presentation of the apparently factual biographical details for each of the women provided at the close of the films. (2004, p. 50)

Mullan visited the archive to view the Delaney footage, long before it became available online through the IFI player.[1] When depositing the collection in the 1990s, Father Delaney's niece, Irene Devitt, signed a donor agreement to allow researchers to view and reuse the film, in the knowledge that in theory it could be used in a production that criticised the Catholic Church. However, on Home Movie Heritage Day in August 2010 in the Irish Film Institute in Dublin, she voiced her discomfort with Mullan's reference to the collection, as she felt it implicated her uncle by showing the filmmaker depicted as a cruel and abusive priest. Although the character in the film did not share the same name, the family believed an implicit connection was made.

Having previously worked with Ken Loach, Mullan cites the director as a key influence on his work. While the two have some similarities in their blended approach to historical characters to highlight injustice, Mullan's borrowing is different from Loach's. Jessica Scarlata (2014) suggests that the scene depicting amateur film enthusiast Father Fitzroy,

> discredits a staple of the traditional archive and a weapon in the arsenal of 'reality' – non-fiction footage – by revealing that what passes for the real is only as reliable as the person behind the camera. (p. 240)

This extends beyond the character of Father Fitzroy to Peter Mullan himself. O'Flynn's insight that Mullan consulted and re-enacted *real* non-fiction footage compounds this claim. Furthermore, Scarlata's (2014) comment that 'the "documentary" footage that Father Fitzroy shoots to prove the Church's benevolence and generosity regarding the Magdalenes is no closer to the truth of what the inmates experienced than the Hollywood production of *The Bells of Saint Mary's* (shown later in the film) is to how the clergy treat their charges' (p. 240), becomes all the more potent in the knowledge that this scene is a reconstruction, almost scene-by-scene, of the footage shot by Father Delaney in the 1930s. This extra element of reality is exactly what upset the priest's family, because comparisons are drawn between the original amateur filmmaker who certainly filmed the Magdalenes, but was not guilty of the abuse perpetrated by the grotesque voyeur depicted in the film.

Crucially,

> other amateur films documenting the everyday lives of such institutions and the orders who ran them have been withheld from programme makers. The owners of rights in these films (that is, the religious orders who made or commissioned the films) have argued that the anonymity of the subjects of the films must be preserved. (O'Flynn, 2012, p. 50)

This embargo leaves many clerical collections under lock and key, at least until a point in the future when they may be revealed to the general public. In the meantime, one can only speculate on what they contain and this leads again to questions of archival power and public access.

There has been a recent controversy in relation to the short film *Detainment* (Vincent Lambe, 2019) billed as 'a true story based on interview transcripts and records'. The film features harrowing depictions of the police interviews with Jon Venables and Robert Thompson, the 10-year-old killers of toddler James Bulger. The film garnered critical acclaim and was shortlisted for the Oscars, but director Vincent Lambe faced huge backlash for not consulting the Bulger family before its release. While offering an apology for this, he has argued that he relied on archival documents so that 'everything in the film is entirely factual and there are no embellishments whatsoever'. Speaking on Today FM, Lambe stated:

> Everything in the film is already public record, from that point of view, you don't need permission to tell somebody something that is already public record, it's been shown in several documentaries already. It's all there, there's nothing I'm changing with it. We wanted to rely solely on the factual material so I think by talking to one family and not another or any of the families, you would, even if it was subconsciously, be putting an opinion in the film. And we didn't want to do that, we wanted to make a film which was entirely factual and based on the evidence that was already there. (Lambe quoted in Stacey, 2018)

Lambe's account is somewhat reductive, failing to acknowledge the editorial choices made in relation to which transcripts were chosen and how these were framed in the narrative. Both of these examples offer evidence of the complexity of archival research for visual media production. While both filmmakers can argue that technically they followed the letter of the law, the reactions of the families associated with these real cases demonstrate that they found the depictions offensive and hurtful. Both films are harrowing; both dramatise real events to raise important questions about a society in which torture and cruelty can be inflicted. The ethical question here relates to whether the audience can trust the filmmaker to handle the material 'appropriately' and 'responsibly' and whether the filmmaker can trust his audience to decode what is presented in a sophisticated and nuanced way.

TRUSTING THE FILMMAKER

Peter Jackson recently courted controversy in some quarters with his manipulation of archive footage in the much-lauded World War I documentary *They shall not grow old* (2018). Refreshing for many commentators was the fact that

> the only narration we hear are the voices of the men who fought the war, so the film is blessedly free of the sapient sounds of experts and academic historians. (Gopnik, 2019)

Instead there is a conspicuous presence of dazzling audiovisual effects: colourisation, digital enhancement (particularly of battle sequences) and perhaps most strikingly, added dialogue based on the expert interpretation of lip readers employed to decipher what soldiers may have been saying in the original footage. Jackson stated that 'with his computing power' he could 'erase the technical limitations of hundred-year-old cinema' (Napper, 2018). And indeed he does, with spectacular results technically far superior than any available to the original filmmakers of the 1910s with their heavy cameras and limited resources. There is something poignant in this perhaps; the original footage is extraordinary in itself *because* it is a hundred years old and bears the progression of time whereas Jackson's digital enhancement has added a layer of futuristic restoration that erases these temporal clues and offers a twenty-first century vantage point on the action.

The film begins with black and white footage of soldiers marching in slow motion accompanied with a series of audio accounts which describe the excitement of war. In most cases, a sense of purpose without regret is articulated by the various interviewees. One voice states: 'it was more like a great big game to be enjoyed, apart from the actual shelling and that sort of thing'. The next scenes, in black and white are transposed onto colourful enlistment posters in an aesthetically pleasant montage accompanied by soldiers recounting the outbreak of war and their decisions to join up. These opening scenes echo the jingoistic nature of the original propaganda-laden footage and advertising campaigns and set the tone for the film to come. After exploring the training exercises undertaken by soldiers, it is only at the moment when the soldiers are shipped out to the Western Front to face real battle that the black and white footage becomes 'colourised'. In a moment a little like Dorothy's arrival in Oz (*The wizard of Oz*, Victor Fleming, 1939), the viewer is momentarily disoriented as archive documentary seems to become wartime narrative. There is of course, a tonal difference, as Gopnik points out between the stunning arrival of colour in *The wizard of Oz* and the grim reality of death in the trenches: 'There, it means the coming of magic; here, it means an entry to Hell' (Gopnik, 2019).

This disorientation corresponds with what Baron defines as 'the archive effect', a destabilising sense that one is watching material that has been lifted from its original context and is to be consumed in remediated form. The effect is eerie and unsettling as the eye struggles to understand the vivid and macabre footage. Key to this disorientation is the persistent use of the close-up, a technique not employed by cameramen in the 1910s. Individual faces are selected from group shots and their grinning visages seem to bend and distort under the pressure of this futuristic gaze. As Laurence Napper suggests 'the effect is to force the authentic footage into a bizarre simulacrum of modern feature film-making conventions, rather than to draw out the authentic nature of what it shows' (Napper, 2018). Approaching the original footage with 'enormous audacity' (Gopnik, 2019), Jackson uses a range of digital effects to repackage the original films for a hypermediated, screen-saturated society.

The viewer will inevitably make comparisons with the aesthetics of productions associated with the Second World War such as *Saving private Ryan* (Steven Spielberg, 1998) and *Band of brothers* (Spielberg et al., 2001) with the slightly bleached colour palette and supposed realistic feeling of participating in the action alongside the soldiers, not to mention Speilberg's previous intermittent use of colour in *Schindler's list* (1993). Jackson's footage is intercut with still images: advertisements, paintings, photographs. There are ghoulish depictions of dead bodies enhanced with splashes of red to 'bloody up' the scenes, but the audio continues to be peppered with reminders of the excitement of war. One interviewee remembers fondly:

> when the war was very active, it was really rather fun to be in the front line. It was not very dangerous, a sort of out of door camping holiday with the boys with a slight spice of danger to make it interesting.

The climax of film slows down the footage and intercuts shots of living soldiers with dead bodies while various interviewees describe their injuries, their fallen comrades and their own narrow escape: 'you were face to face with the stark realisation that this was the end of it' one commentator reflects. Sounds of firing and bombardment are added and intensify as the soldiers' accounts become more dramatic. Some of the former exuberance dissipates as we hear the painful effects of these scenes on the soldiers involved, although we are reminded by the end of this section, 'It was a wonderful feeling about being in a battle and I felt very proud of it'. At the end of the film, the footage returns to its archival self: the colour fades to black and white, frame rates return to their erratic speeds and we listen to soldiers describe their challenging return to life after the war.

The aim is to reconstruct the experience of a soldier for the viewer, but of course, this is impossible. Jackson's film is certainly visceral, but it is haunted by the feeling that in spite of all of its impressive effects, it will not make a twenty-first century audience understand war any better. It is certainly entertainment, but it perhaps shares the curse attributed to the war by one of the final commentators of the pieces: 'history will decide in the end that it is not worthwhile'.

This film certainly marks itself as different to other forms of World War I discourse. In many ways, it bears the stamp of an auteur (eschewing the efforts of the original filmmakers). The interviewees, who contributed oral histories to the BBC and the Imperial War Museum, are listed in the credits, and rightly so.

However, the cameramen who shot much of the footage that was reused, in particular Geoffrey Malins and J. B. McDowell are not credited. Famous during the war for their involvement in the production of *The battle of the Somme* (1916) and *The battle of Ancre* (1917), they faced their own battle with the critics for staging some sequences and producing films deemed too upsetting to be screened in cinemas. One cinema exhibitor sported a sign on his front door announcing 'we are not showing *The battle of the Somme*. This is a place of amusement, not a chamber of horrors' (Chambers, 2012, p. 17). Malins (1993) was proud of the war scenes he filmed, suggesting that his work had an 'astounding realism' (p. 177) and explaining that war films:

> must not leave the public with a bitter taste in their mouth at the end. The film takes you to the grave, but it must not leave you there; it shows you death in all its grim nakedness; but after that it is essential that you should be restored to a sense of cheerfulness and joy. That joy comes of the knowledge that in all this whirlpool of horrors our lads continue to smile the smile of victory. (p. 181)

All this context is lost in Jackson's film, and yet the jingoistic spirit of Malins' work is also evident in Jackson's. In many ways, *They shall not grow old* bears many of the traces of Jackson's other work – it is an epic production replete with spectacular battle scenes and stunning special effects just like the *The lovely bones* (2009) or *The lord of the rings* films (2001–2003). One of the biggest issues for archivists is that Jackson is not working with fantasy here, but with the actuality material that documents significant moments in history.

What Jackson has performed is not restoration, as some archivists have pointed out, it is rather artistic license and creative enhancement to produce what Kasandra O'Connell, Head of the IFI Irish Film Archive, has described as 'a new art form' (*The History Show*, 2019). Archivists, of course, restore footage all the time: every time film runs through a projector it degrades, gathering dust and scratches. Film is often cleaned and respliced; frame rates and under- or over-exposure are often adjusted and colour film is graded in the digitisation process. But this is always done while acknowledging the spirit in which the film was first produced and shown and restoration brings film back as close as possible to the original version exhibited in the cinema. In most archives, this 'restored' version is preserved as a master copy and digital viewing copies are made to allow researchers to view and use the material without further damaging the original film stock. It is worth considering the tension between the archivist and the filmmaker here: archivists seek to preserve and protect film as part of our audiovisual heritage; filmmakers seek to use that preserved resource to entertain and inform. While many archives now prioritise ways in which their collections can be made accessible to the general public, sometimes there are clashes in terms of respecting the source that is used and guiding the audience in how they should 'read' archival images.

TRUSTING THE AUDIENCE

Oscar-winning film producer Lord David Puttnam, known for working on historical films often based on real characters, cites a chilling example of how cinema can encourage an audience to emote in ways never anticipated by the filmmaker. Puttnam's concern about the responsibility of the film industry is well known, having argued 'Movies are powerful. Good or bad, they tinker around inside your brain. They steal up on you in the darkness of the cinema to inform or confirm social attitudes' (Puttnam, 1998, p. 270). *Midnight express* (1978), created by the triumvirate of Alan Parker as director, Oliver Stone as writer, and Puttnam as producer, depicted the story of Billy Hayes, an American college student incarcerated in a Turkish prison after being caught attempting to smuggle drugs out of the country. Puttnam (1998) describes a screening of the film in New York:

> There is a scene in which our 'hero', Billy Hayes, bites off the tongue of a fellow prisoner. When we made the film it seemed as powerful a way as any of illustrating the degradation of the human spirit that's possible under these conditions. But in the cinema it had an effect quite unforeseen – at any rate, unforeseen by me. Some people were so infected by the cliché of brutal retribution for brutal treatment that they leaped to their feet, cheered, and applauded. I was genuinely appalled. Ever since, I've been very cautious about the use of violence in film. (p. 5)

This encapsulates the responsibility of the filmmaker who appropriates 'real' people or events to tread carefully in depicting sensitive material for audiences; the filmmaker acts as a gatekeeper and requires a moral sensibility when dealing with content or contexts that may incite hatred or propose action. Puttnam (1998) goes on to take a Socratic approach in suggesting a personal code of ethics for filmmakers:

> I have come increasingly to reject the idea that filmmakers should in any way conspire to put their work above or outside what they believe to be a decent set of values for their own life, their own family and their future in society. (pp. 5–6)

Alan Gilsenan (2012) echoes these sentiments: 'In the end, ethics is fundamentally about honesty and respect for others, and ethics for [film]making are no different' (p. 64). Filmmaking then, carries with it, an enormous potential in transmitting its chosen message or agenda. Puttnam (1998) articulates the transformative power of film to challenge social injustices in 'a morally responsible community, in which the artist can both function and be encouraged to constitute part of the solution' (p. 6).

Many of the ethical questions considered in this chapter relate to deceased persons: Anne Frank, the numerous victims of the Holocaust, Ayrton Senna, Amy Winehouse, Father Jack Delaney, James Bulger, the soldiers of World War I, and perhaps (or perhaps not) the anonymous woman on holiday, captured in an exhibition in Luxembourg. In examining the examples cited above, we might extrapolate an ethical analysis based on Couldry's trio of ideal values: accuracy, sincerity, and care. In examining Loach's appropriation of the archive, it is significant that he bends historical accuracy by refraining from showing the main news report available covering the Treaty (*Peace Council at the Palace*, 1921), but rather creates his own amalgamation of sources masquerading as a coherent whole. Furthermore, we know that this news would not have been revealed for the first time to an audience in the cinema but rather through word of mouth, letters, or the newspaper. But while Loach plays with historical accuracies, he does achieve a high level of *authencity* in recreating the experience of a 1920s audience and there is most certainly sincerity in his wish to represent the frustration felt in some quarters in reaction to the Anglo-Irish Treaty. There is care taken to represent this visually for the audience, allowing the authority of the archive to break into the drama to move the narrative along efficiently, with due acknowledgement that these events were happening to real people, with devastating effects. In this sense, Loach adheres to two of Couldry's trio of ideal values and substitutes the third for authenticity (rather than accuracy) but the director still faced significant scrutiny from the mainstream media for his perceived political bias and his use of archive served his agenda of depicting the unsatisfactory nature of the peace treaty offered to the Irish. Arguably, this moment is pivotal in the film as is shows the split between the screen (depicting the real Michael Collins as pro-Treaty) and audience (showing fictional protagonist Damien played by Cillian Murphy as anti-Treaty). It is interesting that Loach chooses this moment, where the county is on the brink of civil war, to return to the archive to add 'authentic' exposition to his story. This crucial moment demands a gravitas that forces Loach to return to the 'real' depiction of events held within the archive. In excavating and reworking this material, Loach draws our attention to the fact that he has fabricated this intervention into the narrative in the same way that British propaganda was constructed to portray the Treaty as a success for the Irish and British public. Thus the audience is asked to critique his action, and all accounts of Irish history and politics, as narratives in need of nuanced deconstruction.

Peter Mullan also bends historical accuracy with his blurring of fact and fiction, yet clearly approached the chosen subject matter with sincerity as his main aim was to document and challenge the horrific abuses suffered by Magdalene women whose voices had been silenced. Father Delaney's niece, Irene Devitt, commented on the family's expectation that the Delaney collection would be used in documentaries that were critical of the Catholic Church, but that Peter Mullan's representation of the footage in *The Magdalene sisters* could not have been anticipated. These remarks were made to an audience at *Home Movie Heritage Day* in the Irish Film Institute on 21 August 2010. (The Delaney collection was also the subject of a documentary, *Father Delaney: Silent Witness*, broadcast on RTÉ One at 10.15 p.m. on 24 October 2019.) While there was care shown towards the women depicted, this was not extended to Father Delaney, the priest whose films inspired a striking sequence, causing hurt to Delaney's family who felt there may have been a comparison drawn between the real priest and the grotesque abuser characterised on screen. Equally, Vincent Lambe demonstrates a high level of supposed accuracy in his use of transcripts in *Detainment* and there is sincerity in his attempt to understand what drove two young boys to commit such horrific acts. Arguably, he has not accorded due care to the family of James Bulger in failing to engage in dialogue with them before the film's release.

Jackson pushes the boundaries of accuracy using technology to lift archive footage from the vaults and bring it into a spectacular, dramatic world of special effects, promising an authentic experience of life in the trenches for the viewer. There is personal sincerity here as Jackson cites his grandfather's involvement in the war as his inspiration, yet the overall film seems drastically distant from the original soldiers as the eyes and ears are encouraged to focus on the experience of the film, without thinking too heavily about the complex and nuanced context of war. Little care was taken to acknowledge the original filmmakers or to provide an educational context for the film, although there are now plans to create an educational booklet for use in schools. Kapadia offers an interesting example whereby in one scenario it was easier to follow the values of accuracy, sincerity and care by upholding a strong relationship with the family of Ayrton Senna, whereas this relationship with Amy Winehouse's family broke down, causing a conflict in the film's reception and some degree of distress to her relatives.

We might then suggest that Loach approaches the archive in a more ethical way than Mullen, Lambe, and Jackson and causes least damage to persons still living in doing so. It is also interesting that Loach trusts his audience to decode what is presented to them and thus critique all media representations of conflict. Kapadia offers us a more complicated reading where his approach could be deemed ethical in all three core values in one film concerning a deceased person and somewhat lacking in another. Yet a counterargument here would suggest that it is not Kapadia's ethics that have changed, but rather the external circumstances of production.

APPROACHING THE ARCHIVE AS ACADEMIC CREATIVE PRACTICE

In drawing together this melee of somewhat contradictory examples, it is possible to make some tentative recommendations for researchers who intend to reproduce archival film, particularly those working within the academy. Responsible research practices should invoke Couldry's trio of values, keeping accuracy (or at least authenticity), sincerity, and care at the heart of the research. In doing so, researchers should highlight any sensitivities that may arise in relation to the original conditions of production and any controversy that may occur in the reuse of the source material. In seeking guidance on this, it is imperative to build open and transparent relationships with archivists as the custodians of the material concerned.

Archivists have a wealth of knowledge and know their collections in ways that cannot be captured by catalogues alone. Building a good relationship with an archivist can unleash a range of information that will illuminate the reuse of material in immeasurable ways. Archivists also tend to 'bear in mind the unforeseeable uses of all kinds of footage' (Becker cited in Rao, 2010, p. 109) and often act as a guiding conscience to filmmakers and researchers when advising on the protection of the privacy of the subject of reuse. It is essential, then, that researchers consult with and include archivists in their research process. The Head of the IFI Irish Film Archive Kasandra O'Connell shares her experiences as a trained archivist with students in Critical and Creative Media at NUI Maynooth. Her invaluable input is one which students could not access through academics who have visited, but never worked in an archive, and these students will approach creative recycling with practices that are respectful of the source material as a result. Other institutions would do well to follow this model.

It is also important to build trust with stakeholders and others who might be affected by viewing the completed research outputs. Such consideration for how a film might affect those close to the subject matter or the source material is an essential part of any creative endeavour that seeks to espouse good ethics and particularly relevant for practitioners working in academia. Making informed decisions on the tension between truthful representation and a basic human consideration of the sensitivities of others should be at the core of any ethical production and at the heart of all academic pursuit of new knowledge.

It should also be acknowledged that stakeholders may change their mind about how their footage or personal narratives are used during the course of any project undertaken. As Rao (2010) outlines:

> Artistic works are subject to change during the process of their creation ... [T]he process of filming and carrying out the project may change the filmmaker's understanding of the subject in ways that alter the final product. The consent given by the subject at the time of filming is therefore based on trust in the filmmaker rather than on an informed understanding of the final product. (p. 109)

Filmmakers should thus keep in mind 'the problematic or changeable nature of consent and the differing conditions under which consent may be granted and later change' (Rao, 2010, p. 120).

A sense of the original context of the reused footage should be retained within the production or in contextual supplementary material: Just as 'the archive has an ethical responsibility to inform any viewing audience of relevant context and to encourage an understanding of the peoples and cultures depicted as well as a responsibility to make policy decisions that encourage culture awareness' (Rao, 2010, p. 117) so too should the filmmaker bear this responsibility. This extends to any reuse that directly misleads the audience, rather critique should be invited of both the production and the ways in which it has recycled the archive.

In the days when archives were public institutions affiliated with governments they were governed by what Jacques Derrida described as 'archons'. These citizens 'who held and signified political power were considered to possess the right to make or represent the law' and they also became the guardians of official documents (Derrida, 1995, p. 2). Thus, archives were inevitably associated with power and hierarchy. Those who interpret the archive inherit that hierarchical legacy. If filmmakers remediate archival content for viewers, then is it not incumbent upon them to consider their ethical responsibilities and their own understanding of the archive in doing so? Foucault (1972) suggests that:

> the archive cannot be described in its totality...it emerges in fragments, regions, and levels, more fully, no doubt, and with greater sharpness, the greater the time that separates us from it. (p. 130)

Both theorists acknowledge the potential for greater understanding of the archive over time and yet both suggest that a full understanding may always be tantalisingly out of reach. This is perhaps, also the case in relation to ethical approaches to the archive. Time and distance may bring a clearer critique in relation to what archival material is used and how and yet the notion of an ideal set of ethics for appropriating the archive remains elusive. To return to Alan

Gilsenan, well known for his expert archival appropriations in *Home movie nights* (1996–1998) and *The yellow bittern* (2009) and his remarkable use of transcripts and personal notes to develop his script for *The meeting* (2018), we are reminded 'how utterly subjective, how gloriously ambiguous all this talk of ethics really is ... the very notion of a definite set of [filmmaking] ethics remains as elusive as ever' (Gilsenan, 2012, p. 65).

NOTE

1. Available at: https://ifiplayer.ie/category/delaney/.

ACKNOWLEDGEMENTS

Thanks to Rod Stoneman for his enthusiastic discussions and generous feedback on this chapter and for highlighting many useful examples of complex archival reuse.

REFERENCES

Amy Interview Special. (2015, June 28). *Film 4*. Retrieved from https://www.youtube.com/watch?v=koYpJdNGZ9M

Baron, J. (2014). *The archive effect: Found footage and the audiovisual experience of history*. New York, NY: Routledge.

Benjamin, W. (1935). The work of art in the age of mechanical reproduction. In L. Braudy & M. Cohen (Eds.), *Film theory and criticism* (pp. 731–751). Oxford: Oxford University Press.

Chambers, C. (2012). *Ireland in the newsreels*. Dublin: Irish Academic Press.

Chambers, C. (2018). The Irish question: Newsreels and national identity. In C.Chambers M. Jönsson & R. Vande Winkel (Eds.), *Researching newsreels: Local, national and transnational case studies* (pp. 265–283). New York, NY: Palgrave Macmillan.

Couldry, N., Madianou, M., & Pinchevski, A. (Eds.). (2013). *Ethics of media*. London: Palgrave Macmillan.

Derrida, J. (1995). *Archive fever: A Freudian impression*. Chicago, IL: University of Chicago Press.

Foucault, M. (1972). *The archaeology of knowledge*. London: Tavistock Publications.

Gabbatt, A. (2009). Film footage of Anne Frank posted on YouTube. *The Guardian*, October 2. Retrieved from https://www.theguardian.com/world/2009/oct/02/anne-frank-video-release-youtube

Gilsenan, A. (2012). True North: The ethics of consent. In C. MacKeogh & D. O'Connell (Eds.), *Documentary in a changing state: Ireland since the 1990s (pp. 57–65)*. Cork: Cork University Press.

Gopnik, A. (2019) A few thoughts on the "authenticity" of Peter Jackson's *They shall not grow old. The New Yorker*, January 14. Retrieved from https://www.newyorker.com/news/daily-comment/a-few-thoughts-on-the-authenticity-of-peter-jacksons-they-shall-not-grow-old

Holden, S. (2011). A spectacular rise and fall. *The New York Times*, August 11. Retrieved from https://www.nytimes.com/2011/08/12/movies/senna-documentary-on-racecar-driver-review.html

Interview with the archive researchers on *Diego Maradona*, Asif Kapadia's latest documentary. *Archive Valley*. Retrieved from https://archivevalley.com/tag/archive-footage

Jeffries, S. (2011). The Saturday interview: Asif Kapadia. *The Guardian*, July 9. Retrieved from https://www.theguardian.com/theguardian/2011/jul/09/asif-kapadia-interview-ayrton-senna

Malins, G. (1993). *How I filmed the war*. London: Imperial War Museum.

Napper, L. (2018, October 23). *They shall not grow old* and the elephant in the room. The International Association for Media and History Blog. Retrieved from http://iamhist.net/2018/10/they_shall_not_grow_old/.

O'Connell, K. (2019). The history show. *RTÉ Radio One*, January 13. Retrieved from https://www.rte.ie/radio1/the-history-show/programmes/2019/0113/1022967-the-history-show-sunday-13-january-2019/?clipid=103017723#103017723

O'Flynn, S. (2004). Black and White and Collar Films. Exploring the Irish film archive collections of clerical films. In R. Barton & H. O'Brien (Eds.), *Keeping it real: Irish film and television* (pp. 39–51). London: Wallflower Press.

Puttnam, D. (1998). *Movies and money*. London: Vintage Books.

Rao, N. (2010). Representation and ethics in moving image archives. *The Moving Image*, *10*(2), 104–123.

Saner, E. (2015). Mitch winehouse on *Amy* the film. *The Guardian*, May 1. Retrieved from https://www.theguardian.com/music/2015/may/01/mitch-winehouse-interview-amy-documentary-film.

Scarlata, J. (2014). *Rethinking occupied Ireland: Gender and incarceration in contemporary Irish film*. Syracuse: Syracuse University Press.

Stacey, S. (2018). *Detainment*: The controversial film of Jamie Bulger's killers. *TodayFM*, January 4. Retrieved from https://www.todayfm.com/podcasts/94662/Detainment:-The-Controversial-Film-Of-Jamie-Bulgers-Killers

Walerich, M. (2012). In S. Kmec & V. Thill (Eds.), *Tourists and nomads: Amateur images of migration* (pp. 193–197). Marburg: Jonas Verlag.

Williams, B. (2002). *Truth and Truthfulness*. An Essay in Geneology. Princeton, NY: Princeton University Press.

FILMOGRAPHY

Fleming, V. (1939). *The wizard of Oz*. USA.

Frears, J. (2013). *Philomena*. UK, USA, France.

Gilsenan, A. (1996–1998). *Home movie nights*. RTÉ, Ireland.

Gilsenan, A. (2009). *The yellow bittern*. Ireland, USA.

Gilsenan, A. (2018). *The meeting*. Ireland.

Jackson, P. (2001–2003). *The lord of the rings*. New Zealand, USA.

Jackson, P. (2009). *The lovely bones*. USA, UK, New Zealand.

Jackson, P. (2018). *They shall not grow old*. UK, New Zealand.

Kapadia, A. (2015). *Amy*. UK.

Kapadia, A. (2010). *Senna*. UK, France, USA.

Kapadia, A. (2019). *Diego Maradona*. UK.

Lambe, V. (2018). *Detainment*. Ireland, UK.

Loach, K. (2006). *The wind that shakes the barley*. Ireland, UK.

Malins, G. (1916). *The battle of the Somme*. UK.

Mallins, G., & McDowell, J. B. (1917). *The battle of Ancre*. UK.

McCarey, L. (1945). *The bells of St. Mary's*. USA.

Mullan, P. (2002). *The Magdalene sisters*. Ireland, UK.

Parker, A. (1978). *Midnight express*. UK, USA.

Spielberg, S. (1993). *Schindler's list*. USA.

Spielberg, S. (1998). *Saving private Ryan*. USA.

Spielberg, S.et al (2001). *Band of brothers*. USA.

The following archival material may be viewed via the IFI Irish Film Archive's Player:

Peace Council at the Palace. Topical Budget and British Pathé, 8th December 1921. Retrieved from https://ifiplayer.ie/peace-council-at-the-palace-pathe/ see also http://bufvc.ac.uk/newsonscreen/search/index.php/story/128327

The Father Delaney Collection. Retrieved from https://ifiplayer.ie/category/delaney/

CHAPTER 8

ETHICAL ISSUES IN A PARTICIPATORY PHOTOGRAPHY RESEARCH PROJECT INVOLVING YOUTH WITH REFUGEE EXPERIENCE

Susan M. Brigham and Mohamed Kharbach

ABSTRACT

Photography is used in research because of its appeal for communicating, expressing feelings, sharing experiences, raising new awareness of participants and potential audiences, clarifying social issues, and framing plans for action. Taking and sharing photos has become easier particularly because of ready access to devices with cameras. Yet, using photographs in research can undermine anonymity and confidentiality (Noland, 2006), and unanticipated unauthorised dissemination of digital images raises ethical concerns for researchers using photography in their research methods (Brigham, Baillie Abidi, & Calatayud, 2018). In this chapter, the authors discuss the participatory photography method and provide practical suggestions for carrying out ethical research using participatory photography. The authors highlight the cultural, social, and contextual situatedness of ethics by drawing on our own research project with youth with refugee experience.

Keywords: Participatory photography; dignity; consent; refugees and immigrants; youth; arts-based research

---

Ethics and Integrity in Visual Research Methods:
Advances in Research Ethics and Integrity, Volume 5, 153–170

ISSN: 2398-6018/doi:10.1108/S2398-601820200000005014

INTRODUCTION

Photographs have the power to evoke a profound sense of understanding through empathic experience (Eisner, 2007). Photographic images have brought into sharp focus local and global concerns of environmental devastation, natural disasters, war, famine, migration, gendered violence, military suppression, poverty, oppression of workers, and many other social issues. They have sparked an array of responses, and some have spurred social activism and change. For example, the widely circulated image of three-year-old Aylan Kurdi's body, washed ashore on a beach in Turkey in 2017, generated attention to the humanitarian crisis caused by the war in Syria. Slovic, Västfjäll, Erlandson, and Gregory (2017) assert that this 'iconic photo of a single child had more impact than statistical reports of hundreds of thousands of deaths', adding, 'people who had been unmoved by the relentlessly rising death toll in Syria suddenly appeared to care much more after having seen Aylan's photograph' (p. 640).

Because of the power of photography to arouse emotions, promote deep reflection, and communicate feelings, ideas, and experiences (Brigham, Baillie Abidi, & Calatayud, 2018), researchers have turned to photography as a research method. One such method is participatory photography. Participatory photography offers valuable opportunities to expand the depth of research participants' voices as they share their stories, name their realities, engage in critical dialogue, and promote awareness of their experiences within a group (Brigham, Baillie Abidi, & Calatayud, 2018, p. 104). The photos and the narratives related to the photos lend themselves to creative and engaging knowledge dissemination, increasing the accessibility of the participants' stories, broadening the audience and having potential impact.

While recognising the value of photography in research, we also acknowledge there are complex and challenging ethical considerations involved in the participatory photography research process. Indeed, 'the act of taking pictures in any community is a political act' (Castleden, Garvin, & Huu-ay-aht First Nation, 2008, p. 1396). Ethical research implicates a deliberate intention to do no harm.

In this chapter, we refer to a study we recently completed with colleagues.[1] The purpose was to support youth, who came to the Halifax, Nova Scotia as refugees, in exploring concepts and issues of social justice, social activism, migration, and learning about participatory photography methods. Over a 12-week period, 10 youth between the ages of 16 and 24 with refugee experience participated in a series of workshops where they received basic training in camera operation and photo editing skills, as well as time spent on discussing concepts of social justice and activism and what these terms mean to them. Youth used the skills developed in the workshops to take photos that communicate their lived experiences and the social justice concepts they wished to share. At each session, which lasted about two hours, participants shared and discussed their photos with the group. Discussions were supported by a group facilitator. While we offered basic point and shoot cameras, all participants chose to use their cell phones as cameras. At the end of the 12-week period, a public forum was organised by the participants and research team to showcase the work of participants. The event provided a further opportunity for participants to develop their public speaking and advocacy skills.

The purpose of this chapter is to provide the reader with insight into the methodology and practical suggestions for carrying out ethical research using participatory photography. We start with a brief definition of participatory photography, then we discuss the ethical considerations implicated in four stages, specifically related to ethical practices we encountered in our research study, followed by a conclusion.

PARTICIPATORY PHOTOGRAPHY

Participatory photography is an arts-based method. The essential purposes of arts-based research methods are to:

> enhance understanding of the human condition through alternative (to conventional) processes and representational forms of inquiry, and to reach multiple audiences by making scholarship more accessible. The methodology infuses the languages, processes, and forms of literary, visual, and performing arts with the expansive possibilities of scholarly inquiry for purposes of advancing knowledge. (Cole & Knowles, 2007, p. 59)

For us, participatory photography is an umbrella concept encompassing different collaborative forms of photography-based methods. In general terms, participatory photography is a collaborative research method in which research participants are actively involved in taking photographs to document their lived experiences, tell their stories, explore community needs, and create awareness of their experiences and circumstances within a group, and possibly with a wider audience. The method aims to encourage self-awareness and group discussion, and develop collective knowledge. It is a valuable research method for researchers working with marginalised groups (Gotschi, Delve, & Freyer, 2009; Prins, 2010), including refugee and im/migrant youth (Brigham, 2015; Brigham, Baillie Abidi, & Calatayud, 2018; Brigham, Baillie Abidi, & Zhang, 2018; Fassetta, 2016; Robertson, Gifford, McMichael, & Correa-Velez, 2016), people with disabilities (Jurkowski, 2008; Newman et al., 2009; Povee, Bishop, & Roberts, 2014), war affected youth (Denov, Doucet, & Kamara, 2012), people with early-stage Alzheimer disease (Wiersma, 2011), women experiencing chemotherapy (Frith & Harcourt, 2007), children living in orphanages (Johnson, 2011), Indigenous communities (Castleden et al., 2008), children with autism (Carnahan, 2006), and people who are homeless (Hodgetts, Radley, Chamberlain, & Hodgetts, 2007; Radley, Hodgetts, & Cullen, 2005; Wang, Cash, & Powers, 2000). The method allows for participants to communicate stories without relying solely on words. It is an appropriate method for those whose first language is not the same as the other research participants and/or researchers. The research participants can be engaged in the research 'in a meaningful way, either as the subjects of the research, co-researchers, or as researchers of their own experiences' (Eliadou, 2015, cited in Barromi Perlman, 2016, p. 6). It enables the participants to have control over the dissemination of the photographs.

The premises behind using participatory photography in research are that photos can:

(a) prod and sharpen memory, reduce misunderstandings, and stimulate emotional storytelling (Collier, 1957, p. 858);
(b) raise new awareness of research participants' consciousness of their social existence (Harper, 2002, p. 21);
(c) help clarify a social issue and frame plans for action (Singhal, Harter, Chitnis, & Sharma, 2007, p. 216); and
(d) potentially reach a wide audience including policy-makers and the public in general.

We do not use participatory photography and photovoice interchangeably as some scholars do (e.g. Johnson, 2011). Photovoice is a specific method developed by Caroline C. Wang and Mary Ann Burris (1994), and while it is like participatory photography because it is participatory and the participants are co-researchers and have control over their photos and narratives, photovoice is committed to system-level change. We preferred to have participants decide if and how they would reach policy-makers and others. Our concern was that if we had declared that there would be a system-level change as a result of the project and none occurred, our research participants would have developed a lack of confidence and trust in the research team and the project. Ethical considerations are at the research preparation stage.

Ethical considerations related to 'the protection of subjects from harm, the right to privacy, the notion of informed consent, and the issue of deception' (Merriam, 2009, p. 230) have to be addressed at the preparation stage, that is, when designing the research; before recruitment. In our study, preparation was not a simple matter for several reasons.

First, as with any research, researchers must reckon with issues of power, trust, and ownership (Castleden et al., 2008). The benefit of participatory photography is that it is participatory and the participants are in effect co-researchers who have control over their representations in the study (Brigham, Baillie Abidi, & Calatayud, 2018, p. 105) thereby lessening the power imbalance between researchers and participants. However, as researchers, we are

> privileged, highly educated, and professionally trained … [who], as social actors, may … reproduce unequal social and power relations, such as race, gender, and class relations between themselves and participants within the social institution and a larger global and neo-liberal context. (Zhu, 2019, p.64)

We had to avoid a 'fixed method' that reflects an 'academic trend of doing "parachute" research' (Castleden et al., 2008, p. 1401). Parachute research is when a researcher parachutes into a community for a short time, without taking the required time for the community to know them for the researcher to get to know the community in an effort to build trust.

Further, sustained participation by a group over an extended period of time is required to develop a sense of community between the participants and researchers (Brigham, Baillie Abidi, & Zhang, 2018). We discuss the time-intensive aspect of our research below and elaborate on how that may be an ethical concern. In our study, our research team included two team members who work with the Immigrant Services Association of Nova Scotia (ISANS), which has youth-focussed programmes. One of these team members worked closely with youth and had an established relationship through the programmes with some of the youth who subsequently participated in our project. Additionally, the first author has been working in the Halifax community with refugee and immigrant families for over 15 years and is familiar to several refugee and immigrant families. These relationships and positive reputations in the community helped the team members to build trust over time. Opportunities for participants and the research team members to get to know one another were achieved through icebreaker activities during every session.

Researchers must become more conscious of power imbalances between themselves and the participants and within communities and between potential participants. Gotschi et al. (2009) who involved women and men farmers in Mozambique in their photography project found that the women participants, who were explicitly allocated the cameras and trained in their use, had their cameras taken away by their husbands or male leaders. The resulting photos in the project were predominantly from a male perspective. Further, Gotschi et al. reported that a male leader used the camera to 'increase his reputation by promising people to get them their pictures' but the camera had a limited number of shots, so not everyone got their picture taken resulting in the repeating of the first stage of the project. These examples point to ethical dilemmas, including: the women's possession of the borrowed cameras created unanticipated impacts on relationships in their families and in the community at large, and the resulting 'data' (i.e. photographs) had a limited women's perspective, which was the main research focus. This seemed to be partly because the participants lacked clarity for the research purpose (suggesting a communication problem between the researchers and participants), the researchers were not fully aware of the gendered power differentials in the community, and the community at large did not have sufficient 'buy in' for the research project. In that research project, participatory photography may not have been the most appropriate method.

In our study, potential participants met one-to-one with two team members to have the project explained to them and where they could ask clarifying questions, an information letter was given to each participant, and more than twice participants reviewed the consent form with research team members. Additionally, even though participants were offered cameras and were trained by a professional photographer on how to use them, they preferred to use their own cellphones. This proved helpful as the ubiquity of cellphones reduces the visibility of these devices within community and lessens what Sontag (1977) refers to as the intrusive feel that a traditional camera sometimes arouses. As such, our participants felt a higher degree of ease and comfort as they went about capturing their individual experiences and photographing things meaningful to them. Furthermore, the first author had worked with the community on two previous participatory photography projects that had culminated in several public events, which some of the participants had heard about. Hence, the photography research process was somewhat familiar to some of our youth participants.

Second, the participatory photography method often leads to culminating events intended to share participants' perspectives with wider audiences and raise awareness of issues with the public and policy-makers (such as photography displays or community shows). Therefore, research projects using participatory photography will usually entail involving established community partners who have networks in the community and long-term connections with policy-makers/ people in positions of power early on in the research stage. Community partnerships also allow for a broader reach for recruitment and they can give insights into wording and appropriate phrasing in the recruitment materials. For example, before recruitment, our research team (which consisted of two people from an established immigrant settlement association, a member of a youth-focussed arts-based organisation, two academic researchers, and several research assistants) met to discuss the appropriate terminology for refugee youth – our target participant group. After discussion, we settled on 'youth with refugee experience' because some youth with whom the team members had worked with in the past found the term 'refugee' a stigmatising and exclusionary social label.

Third, if there are language proficiency challenges or a lack of a common language among the participants and researchers, it may be necessary to have recruitment tools in several languages and have researchers meet with potential participants one to one to assess their needs and language skills. Strawn and Monama (2012) who worked with women in Soweto, South Africa noted that they involved others to discuss 'the informed consent with participants in their mother tongues' (p. 543). However, the use of interpreters or translators in research can pose challenging ethical dilemma especially if translators are not closely connected to the research project or when translators and research participants belong to what they may perceive as antagonistic ethnic or religious groups. While we did not have to rely on the services of an interpreter because all of our participants were required to have fluency in English in order to participate, some studies demonstrate how the ethicality of research was seriously undermined by the use of interpreters (see Halilovich, 2013; Hopkins, 2008; Kabranian-Melkonian, 2015; Mackenzie & McDowell, 2007; Smith, 2009). Also, during our initial meetings with participants, we took the time to make sure they understood the ethical complications of visual research and that they were aware of the requirements the project demands in terms of effort and time. These initial meetings were also an opportunity to gauge participants' commitment to the project since, as mentioned above, the method requires a significant investment of time.

Fourth, if the research participants are minors, as was the case for some of the participants in our project, it is important to recognise that youth 'are dependent on parents or other adult guardians and lack a voting voice' (Strack, Magill, & McDonagh, 2004, p. 56). Therefore, we, the research team, took a lead in launching the process by drawing on our resources and networks. Specifically, we involved community partners: Youth Art Connection (YAC, 2019) which has established community relationships and networks and whose mission is to work with youth of all backgrounds to grow a successful career and life and the

Immigrant Settlement Association of Nova Scotia, which is committed to helping newcomers build a future where they belong and grow (ISANS, 2019).

Participatory photography often embraces concepts of empowerment, collaboration, and community, which have been themselves the subject of critique. For instance, Hayward et al. (2004, cited in Denison & Stillman, 2012, p. 1040) criticises 'a mythologising of the power of participatory methodologies to accomplish problem solving, emancipation or empowerment'. We agree with Strack et al. (2004) who, although are referring to photovoice, make an important point for participatory photography in general. They advise 'engagement in a photovoice project will not lead to a complete state of empowerment' and allowing youth to believe that 'could leave participants feeling more hopeless and unempowered than when they started the program' (p. 57). Participatory photography methods often aim at social change and empowerment, yet this can sometimes be challenging, especially given the intangible nature of social change and empowerment. Duffy (2011), citing numerous scholars, touches on the polysemic nature of the concept of empowerment, arguing that it can mean learning a new skill, taking part in political action, and gaining critical awareness of surrounding social concerns, among other things. As Duffy (2011) states, 'empowerment as a process or outcome may not always be present or apparent' (p. 3) and its meaning is contextually dependent. Participants who had worked diligently to generate photographs depicting their societal concerns may expect a higher level of engagement from policy-makers and decision takers only to be disappointed at the lack of responsiveness from these officials. For instance, after a photo exhibition in which they engaged in discussions with different policy figures about methods for social change, participants in Denov et al.'s (2012) project felt that

> these discussions ultimately failed to yield concrete changes for the youth [which left them] and [the] research team with concerns about the feasibility of photovoice as a truly viable tool for social change. (p. 130)

A similar concern regarding the social action component is voiced by Johnson (2016), who warned against 'raising false hopes or unrealistic expectations amongst the participants of photovoice projects who are positioned to be the champions for social change in their communities' (p. 799). Instead, he recommended that researchers explain to their participants 'the project has the potential to be policy informing rather than policy changing' (p. 799).

For this reason, we recommend that in the early planning stage, researchers must seek the support and interest of community and policy-makers, a point raised above. While we are referring to youth, this is equally important for adult research participants. Specifically, we involved a community partner (Youth Art Connection) who has established relationships and networks with youth of all backgrounds (YAC, 2019). Additionally, when working with youth, there may be a need to meet with parents or guardians about the project, as they would be required to give consent for their children to be involved. In our project, we did not meet with parents; rather our consent process involved the following process: two team members met with each participant who expressed an interest in the research. Once we got a sense of the participants' abilities to communicate in English (which was a criterion for participating) and their level of commitment, we discussed the consent form and then invited them to meet with the full group of 10 youth. At the first group meeting, participants brought their signed consent forms, which for minors included the signature of their parents. At that time, they met their peers and the whole research team, as well as the counsellor who was available at all times should any youth require professional psychological support. We also went over the consent form again.

Fifth, as the method involves reflecting on and discussing personal experiences and producing photos and stories based on personal experiences, a professional counsellor is often required by university ethics review boards to be available to the participants. Participants in our project had arrived in Canada as refugees; they are from different cultural backgrounds and some had experienced significant trauma. It was essential that our research team consisted of leaders who had experience and training in dealing with these considerations. Before beginning the research sessions, we hired a professional social worker with experience in working with youth who have experienced trauma, who herself is a racial minority, to be available at all times to provide our participants with professional support when needed. Like the research team, the social worker needed to establish trust with the participants over time.

Once the research had been planned, an ethics review had been approved, and the participants were recruited and had consented to participate, we began the training stage.

ETHICAL CONSIDERATIONS AT THE RESEARCH PARTICIPANT TRAINING STAGE

The training of the participants is a critical step, which involves discussing power dynamics, the use of cameras in public places, and ethical issues, including interactions within the group. For example, in our project, to create a safe environment and to ensure participants were emotionally protected, participants together developed community standards, which they discussed and added to each session (see Fig. 1). This included keeping one another's stories in confidence, communicating respectfully and being supportive of one another. It is important to refer to the community standards at each research session and invite further discussion about the standards each time.

Since the participants in participatory photography research projects are taking photos on their own outside of the research sessions, there is a need to plan how participants will describe the purpose of taking photos in the community and for obtaining consent from people who are identifiable in the participants' photos. In our project, Author 1 developed a short consent form that each participant would need to have signed by anyone who appeared in the photos (and if the identifiable subject in the photo was a child, parental/guardian consent was required). In the training, we asked participants to be prepared to explain what they were doing when taking photos in public places, to ask for permission

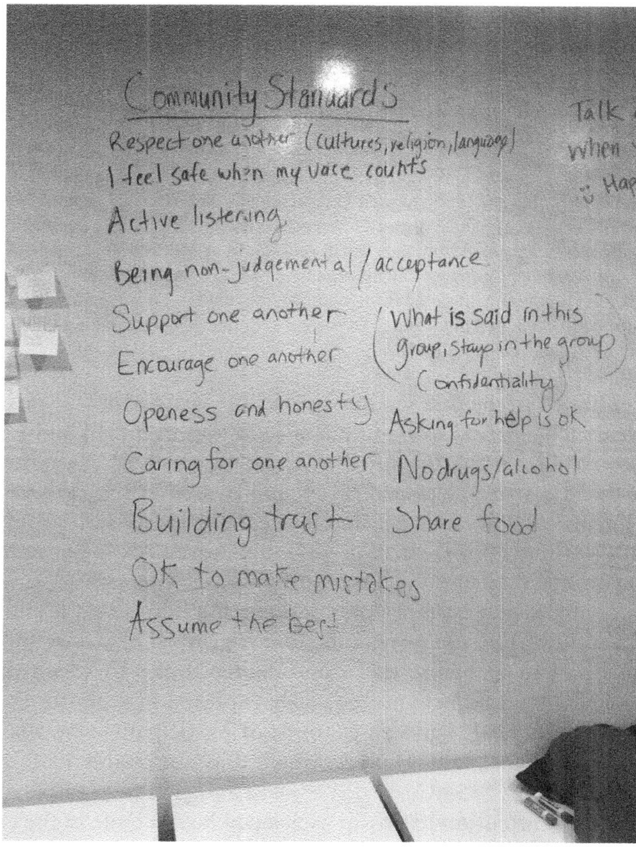

Fig. 1 Community Standards Developed by the Research Participants.

before taking photos of people, to always consider people's safety, and if certain people do not want their photo taken, respect their decision. To help youth feel more comfortable with the process of negotiating social interactions and to practice alternative actions, participants could use role-playing and modelling (Prins, 2010; Strack et al., 2004).

Training also includes training in camera operation, photography skills development, photo editing skills, and safety. The data collecting stage is next, where participants begin taking photos.

ETHICAL CONSIDERATIONS AT THE DATA COLLECTION STAGE

Taking photographs can be an intrusive activity. Indeed, photography is a 'culturally embedded technology of power' (p. 441), which can 'operate as a technology of surveillance that breeds distrust and facilitates social control' (Prins, 2010, p. 4).

Prins (2010) gives examples from her participatory photography study in El Salvador, in which her participants experienced suspicion, hostility, ridicule, and embarrassment as they took photos in the community. Prins (2010) explains:

> [Research participant, Esmeraldaa] went to the cañal [sugar cane field] to take a picture of it and the owner came up and asked her what she was doing. She explained. He asked who was in charge of this project and she gave him my name. He said that I was telling her to take pictures of it so that I could come and 'darle fuego' [set it on fire]. Then Esmeralda went to the school to take a picture of it and the [principal] and a few teachers ended up talking to her and asking why she was taking pictures there. Again, they asked who was in charge and she gave them my name and mentioned Alfalit [the nongovernment organization]. 'If one child disappears, we're going to go look for her [author]' [translated from Spanish], they said... Then Esmeralda went to take a picture of a large [farm machine] and ...the owner said that I was having her take pictures so that I could come and steal it .... (p. 434)

As we mentioned in the first section, preparing participants for possible negative responses, such as criticisms and hostility, is important. In addition, regular opportunities to discuss participants' needs, concerns and emerging problems throughout the study are helpful. Prins reminds us that the camera is not 'an accultural, intrinsically liberating technology that produces similar results in any social cultural setting' (Prins, 2010, p. 427). She advocates for a 'judicious' and socio-culturally informed approach to photography.

The beauty of the participatory photography method is that it allows participants to express ideas that are not limited or constrained by words and language, yet it is important to recognise that non-tangible topics to which participants may wish to bring attention cannot be easily captured in a photo. For instance, Strawn and Monama (2012) give an example of participants who shared a story of sexual harassment in the workplace, which 'did not readily lend itself to the photographic image' (p. 545). Castleden et al. (2008) give an example of a participant's wish to draw attention to gossip as a social health issue in the community. 'Her solution was to photograph the word spelled out on a Scrabble game board' (p. 1402). In my own [Author 1's] previous study, a participant wanted to convey issues related to her identity through poetry, so she wrote the poem, typed it up and took a photo of the written poem. These examples show the creativity of participants in trying to convey ideas that do not lend themselves easily to photographic images. They also suggest that in some cases, a process that includes storytelling or reflective writing that helps to capture narratives and ideas beyond or outside of photographs contributes to maintaining the integrity of the participants' stories. A method that limits the research participants' ability to share their stories in a fulsome way becomes an ethical concern in that participants' voices are partially or fully restricted unnecessarily.

Gotschi et al. (2009) reported that in their study with farmers in Mozambique their participants wanted to photograph what they do at harvesting time, but it was not harvesting season. Therefore, the research participants involved the community in staging photographs where they posed with equipment and props to simulate the activity. The other group of participants in their project simply waited for harvest season to arrive to take photos thereby delaying the project by several months. We suggest that there is no ethical issue with simulation. We take the view that it is the participants' choice about what they want to depict in their photographs and how. However, this example is a reminder that researchers must be flexible and patient, as not all critical activities or events that are important to participants will coincide with the researchers' availability. Researchers would benefit from planning their projects including the time frame and duration for the project with communities.

A challenge for this type of research project, which is often time-intensive, is sporadic attendance, resulting in a small core group of participants that stays involved in the project from beginning to end. The amount of time required of participants is usually significant and as noted above, in order to build a sense of trust and community a long period of time is required for the project. Moreover, many participants who are marginalised are experiencing multiple barriers and demands, leaving them little time for time-intensive projects like participatory photography projects. In our project, we learned of the youths' multiple responsibilities that included: caring for younger siblings and elders, acting as translators for family members at appointments, doing paperwork (such as paying bills and filling in various types of forms), helping family members with school work, running errands, and doing their own school work, paid employment, and social activities. We provided incentives such as the use of a camera for the duration of the study and receiving developed photographs as well as electronic copies of photographs on flash drives at no cost to them. We also compensated participants with gift cards, and provided refreshments and public transportation passes. Yet the reality was that participants could not always participate despite their best intentions. As mentioned above, we interviewed the interested participants before they formally agreed to participate to gauge their interest and availability for the project and we tried to be as flexible as possible with meeting times, yet within the first four weeks of the project we had a number of participants reduce their participation and a few withdrew. This may become an ethical issue as the group loses its sense of cohesiveness and those unable to regularly participate may feel less included when they return to the group after an extended absence. Further, as Bukowski and Buetow (2011) explain, if all their participants (women who are homeless) could have come together as a group it would have 'increased the ability of the women to advocate collectively for what they wanted' (p. 744). Additionally, had their participants been involved over a longer time period there could have been a better focus on the participants' changing situations and a better understanding of why they experienced changes. Ethically, even while participants may be foregoing income in order to participate in the research project, compensation/remuneration must not be excessive as it may impact the participants' sense of free choice to participate.

Financial resources can be limiting factors in the use of photography in research (Coles-Ritchie, Monson, & Moses, 2015). To cover the costs of cameras and other materials required for the conduct of photography-based research projects, researchers have to search for and secure funds, grants, or donations from third party stakeholders. For instance, in their photovoice study with social work students and the lack of sufficient funding, Mulder and Dull (2014) were not able to provide digital cameras to their participants and, instead, asked them to use their own smartphones or any other camera device they had. In our project, we gave the participants the choice of using their cellphone cameras or basic point and shoot cameras. The youth chose to use their own devices. However, we found that using these devices impacted the ability to share photographs as planned because the lower resolution did not allow us to enlarge the participants' photos for a photo display.

ETHICAL CONSIDERATIONS AT THE ANALYSIS AND DISSEMINATION STAGE

In our study, the analysis stage mainly occurred with the participants as they prepared for the dissemination of the project. During group dialogues with facilitators throughout all the sessions, each participant had several opportunities to share and talk about the meaning of their photos and about social justice, which was the overarching theme of our study. As each participant shared, other participants and facilitators asked questions and made comments (e.g. 'That reminds me of a picture I took', or 'That's the same in my culture!'). Participants also chose, arranged, and grouped their own and/or other participants' photos to tell a story or define a theme. In this way, participants were engaged in thematic analysis on both an individual and group basis. Through storytelling, the participants dug into the complexity of themes. For example, while several photos could be grouped as images of nature, through discussion, participants elaborated on how some of these pictures connected to war, food insecurity, gender inequality, environmental destruction, and loneliness. The participants' captioning of the photos for the public event, their public speaking at the event, as well as their songs, dances, and poems reflected their analysis of their photographs and experiences.

At the dissemination stage, maintaining the rights and dignity of the participants, their photographs and any identifiable subject in the photographs is complex. Essentially, at this stage the researcher must consider, 'Is my representation of the participants' stories respectful?' (Wiersma, 2011, p. 213). A strength of the participatory photography approach is that it empowers participants and enables them to be actively involved throughout the research process, including the dissemination and publication of results. Because participants take photos they personally choose to share and usually are involved in deciding what, where, and how to share publically, 'sensitivity to the dignity of participants is implied ... [provided] that they are aware of the subtleties of how images are interpreted and used' (Langmann & Pick, 2014, p. 711). Group dialogue, with facilitation by the research team, helps to draw attention to subtleties. However, dignity is a relative social construct that is context dependent and what might be considered dignified 'in one culture could be considered an indignity in another' (p. 713; see also Lickiss, 2007, p. 29). For instance, in some communities in the Middle East and North Africa, photographing women, even when they may provide consent, can be considered an act of indignity, one that hurts the collective pride of the community. Therefore, researchers and participants need to develop a heightened sense of awareness and sensitivity towards different cultural norms, societal traditions and values in the research context. Protecting the dignity of research participants means, among other things, respecting the participants' culture and acting in culturally appropriate ways. It also means doing research in a way that does not 'demean or reduce the person it involves' (p. 713). To this end, Langmann and Pick suggest that researchers apply what they call a dignity-in-context approach that deals with dignity issues in their situational context. Such an approach has the potential to sensitise the researchers to 'the relativistic nature of social and cultural norms' (p. 713). In our project, all photographs that were shared publically, including those online (social media and websites) were contextualised with descriptions about the purpose of the project and a caption provided by the participant.

Some photography research projects can pose serious risks to the safety of participants. For instance, working on social issues related to violence, drugs, prostitution, gangs and the like can place photographers in dangerous situations akin to the danger photojournalists face when doing investigative work or when covering wars (Peabody, 2013). Researchers and communities must assess the risks and be clear with participants about all potential risks. Similarly, the safety of photographic subjects should be taken into consideration when taking pictures of vulnerable populations and marginalised groups (Peabody, 2013). In some communities, publishing photographs of disempowered groups, such as those from the lesbian, gay, bi-, trans, two-spirit communities might jeopardise their physical safety. Hence, the importance of researchers being aware of the religious and cultural sensitivities underlying their research contexts.

One school of thought about protecting the dignity of participants is to modify the photographic images using digital technology to blur faces and disguise identifiable features in photographs. However, this practice has been critiqued by several scholars such as Susan Sontag, Abigail Solomon-Godeau, and Martha Rosler who, in the words of Choi (2018), argue that photography, 'despite its inherent mechanical objectivity, manipulates, distorts, and thus re-victimises the subject in its pitiless formal attention' (p. 99). The idea of distortion and manipulation can be regarded as a form of identity distortion and objectification of participants (Langmann & Pick, 2014). For Close (2007),

> the original, unaltered image is the one on which interpretation of data rests and to present a defaced image would be no different from presenting interview data whose language, grammar or syntax had been altered. (p. 30)

Likewise, Pauwels (2008) argues that any tinkering with the legibility of images in order to preserve anonymity might affect the communicative strength of photographs and might result in the lost of important data (e.g. contextual background information, non-verbal information such as facial expressions and body language). Banks (2001 cited in Close, 2007) suggests that such 'fuzzy-face effect' in photographs can carry connotations of criminality especially in the Western context (p. 30). Moreover, complete anonymity of research participants using disguised images is difficult to achieve (Pauwels, 2008) particularly if these images 'can reveal important information that text or word-based methods cannot' (Clark, Prosser, & Wiles, 2010. p. 86). In our study, during the editing stage, one participant chose to crop the image to remove the image of a person who had not given consent to be photographed but other than that one instance we did not have any need to manipulate or distort photographs to protect confidentiality.

Given the rise in social media and the ubiquitous nature of taking photos that can be uploaded and shared in an instant, there is an increased chance of unanticipated and unauthorised dissemination of digital images from the research. Posting photos to a website or social media site increases the exposure of participants' stories by reaching a wider audience thereby also extending the potential impact of the research. However, it also means those photos are no longer under the control of the participants or researchers. Neither the participants nor researchers would have control over how the photos are used, manipulated, captioned, contextualised, interpreted, re-posted, and whether the photos will generate positive or negative impressions (Clark et al., 2010; Langmann & Pick, 2014). This may lead to psychological harm (including feeling demeaned, embarrassed, worried, or upset) for the participants and possibly to others by extension (such as family members or any subject in the participants' photos). Addressing this ethical concern requires researchers to be clear with participants about this loss of control in perpetuity and having discussions about the risks over the duration of the project, not just at the dissemination stage. It is also important to make clear to participants that social media sites store information on US-based servers, making the content subject to US laws. Depending on the researchers and/or the institutional ethics review board, researchers may require a separate consent form whereby participants consent to having their photos uploaded to social media sites, specifying how the photo will be presented. Participants should expect that researchers will make every effort to ensure no sensitive, personal information is included in photos and stories that are posted to these sites, and that all participants will be given the opportunity to review and approve stories before they are posted.

As a semiotic form of meaning making, photographs can be ambiguous and sometimes hard to understand. This is mostly due to their context dependence and to the fact that they can be obscurely polysemic (Peck, 2016). Wang and Burris (1997) add that while photographs are easy to collect they are 'difficult to analyze and summarise because they yield an abundance of complex data that can be difficult to digest' (p. 375). Related to this is the unpredictability of the impact they can have on the viewer. For instance, in his photovoice project with youth from violence-affected communities in Kenya, Baú (2015) together with the rest of the research team decided not to share participant-generated photos for fear of causing unpredictable impact on members of the community. Participatory photography researchers recognise that photographs do not represent the world objectively. Photos are social constructs whose significance resides in 'the way the people involved with them understand them, use them, and thereby attribute meaning to them' (Becker, 1998, p. 74). Photos are shaped and controlled by contextual social, cultural, and political elements. They get meaning (and they can have multiple meanings) from the conditions surrounding the making of the photographs and the context in which photographs are viewed (Templin, 1982 cited in Adelman, 1998). Photographs have meanings, which can be inferred from the political, cultural, and social environments where the image was taken and viewed. In this way, photos act not just on mechanical and cognitive levels, but also on an emotional level. From a participatory photography researcher's perspective, this is a powerful aspect of photography that helps participants visualise and share their experiences, thoughts, knowledge, and beliefs with audiences.

CONCLUSION

Ethical research practice is about producing credible and valid knowledge in an ethical way. Ethics in photography-based research is a field of landmines. Researchers are required to exert scrupulous attention and caution in addressing all the ethical issues that might potentially compromise the safety of participants and undermine the validity and credibility of the whole research. The researcher should develop an acute sense of predictability and be able to deal with ethical problems as they unfold during the research process. If anything, the discussion of the various ethical issues covered in this chapter brings to the surface the primary fact that ethical conduct of research requires Riessman's (2005) 'ethics-in-context approach' (p. 473). Such an approach highlights the cultural, social, and contextual situatedness of ethics and requires visual researchers to address their ethical concerns in the light of the immediate context of their research. Clark et al. (2010) call it 'situated visual ethics' (p. 81) arguing that 'research ethics are contested, contextual, dynamic, and … best understood in real, concrete, everyday situations' (p. 82). Ethical research is relativistic and situated in nature and no 'one-size-fits-all ethical policy will emerge for visual research, and indeed, perhaps nor should it' (p. 89).

Besides ensuring the safety of participants and the integrity of the research process and always in the context of addressing ethical considerations, real or potential, photography researchers must strive for a research process where relations of power between the researcher and participants are levelled to the maximum. There is no such thing as a complete power-free relation between researcher and participants (Allen, 2012). One way to do this is through incorporating a participatory and dialogic approach; one that empowers participants, values their voices, and highlights their emic perspective.

We assert that participatory photography is a robust research method with huge potential. We believe the advantages outweigh the limitations for research such as our project. We hope that this chapter demonstrates the complexity of ethical issues in this research method and provides insights for researchers interested in using this method and for researchers who have already had experience with this method.

NOTE

1. The team consisted of: Oladayo Afolabi, Research Assistant; Nabiha Atallah, ISANS; Susan Brigham, MSVU; Simone Chia-Kangata, CYRRC; Louise Hanavan, Project Coordinator; Mohamed Kharbach, Research Assistant; April Mandrona, NSCAD; Hilary Thorne, ISANS; and Ryan Veltmeyer, YAC.

REFERENCES

Adelman, C. (1998). Photocontext. In J. Prosser, (Ed.) *Image-based research: A sourcebook for qualitative researchers* (pp. 131–142). London: Falmer Press.

Allen, Q. (2012). Photographs and stories: Ethics, benefits and dilemmas of using participant photography with Black middle-class male youth. *Qualitative Research, 12*(4), 443–458.

Banks, M. (2001). Visual research methods. Social Research Update, *11*, 1–6. Retrieved from http://sru.soc.surrey.ac.uk/SRU11/SRU11.html

Barromi Perlman, E. (2016). Using visual reflective diaries of photographs of school buildings as a tool for empowering students in teacher training. *International Journal of Qualitative Methods, 15*, 1–15. doi:10.1177/1609406916653760

Baú, V. (2015). Participatory photography for peace: Using images to open up dialogue after violence. *Journal of Peacebuilding & Development, 10*(3), 74–88. doi:10.1080/15423166.2015.1050798

Becker, H. C. (1998). Visual sociology, documentary photography, and photojournalism: It's (almost) all a matter of context. In I. Prosser (Ed.), Image-based research: A sourcebook for qualitative researchers (pp. 47–85). London: Falmer Press.

Brigham, S. (2015). Exploring adult learning spaces for refugee and immigrant learners in Canada through photography. In *Standing conference on university teaching and research in the education of adults (SCUTREA)*, Weetwood Hall Conference Centre, Leeds, England, UK. 7–9 July (pp. 60–67).

Brigham, S., Baillie Abidi, C., & Calatayud, S. (2018). Migrant women learning and teaching through participatory photography. *The Canadian Journal for the Study of Adult Education, 30*(2), 101–114.

Brigham, S. M., Baillie Abidi, C., & Zhang, Y. (2018). What participatory photography can tell us about immigrant and refugee women's learning in Atlantic Canada. *International Journal of Lifelong Education, 37*(2), 234–254. doi:10.1080/02601370.2017.1422044.

Bukowski, K., & Buetow, S. (2011). Making the invisible visible: A photovoice exploration of homeless women's health and lives in central Auckland. *Social Science & Medicine, 72*, 739–746.

Carnahan, C. (2006). Photovoice: Engaging children with autism and their teachers. *Teaching Exceptional Children, 39*(2), 44–50.

Castleden, H., Garvin, T., & Huu-ay-aht First Nation. (2008). Modifying photovoice for community-based participatory indigenous research. *Social Science & Medicine, 66*(6), 1393–1405. doi:10.1016/j.socscimed.2007.11.030

Choi, J. (2018). Photo-Graphy, "The shadow in the cave," its pain and love: The ethics of photography. *Photographies, 11*(1), 95–111. doi:10.1080/17540763.2017.1399286

Clark, A., Prosser, J., & Wiles, R. (2010). Ethical issues in image-based research. *Arts & Health, 2*, 81–93.

Close, H. (2007). The use of photography as a qualitative research tool. *Nurse Researcher, 15*(1), 27–36.

Collier, J. (1957). Photography in anthropology: A report on two experiments. *American Anthropologist, 59*, 843–859.

Cole, A., & Knowles, J. (2007). *Handbook of the arts in qualitative social science research*. Thousand Oaks, CA: Sage Publications.

Coles-Ritchie, M., Monson, B., & Moses, C. (2015). Drawing on dynamic local knowledge through student-generated photography. *Equity & Excellence in Education, 48*(2), 266–282.

Denison, T., & Stillman, L. (2012). Academic and ethical challenges in participatory models of community research. *Information, Communication & Society, 15*(7), 1037–1054. doi:10.1080/1369118X.2012.656138

Denov, M., Doucet, D, & Kamara, A. (2012). Engaging war affected youth through photography: Photovoice with former child soldiers in Sierra Leone. *Intervention, 10*(2), 117–133.

Duffy, L. (2011). "Step-by-step we are stronger": Women's empowerment through photovoice. *Journal of Community Health Nursing, 28*, 105–116.

Eisner, E. (2007). Art and knowledge. In J. Knowles & A. L. Cole (Eds.), *Handbook of the arts in qualitative research: Perspectives, methodologies, examples, and issues* (pp. 3–12). Thousand Oaks, CA: SAGE Publications, Inc.

Emmison, M., Smith, P., & Mayall, M. (2012). Researching the visual (2nd ed.). Los Angeles, CA: SAGE Publications Inc.

Fassetta, G. (2016). Using photography in research with young migrants: addressing questions of visibility, movement and personal spaces. *Children's Geographies*, *14*(6), 701–715. doi:10.1080/14733285.2016.1190811

Frith, H., & Harcourt, D. (2007). Using photographs to capture women's experience of chemotherapy: Reflecting on the method. *Qualitative Health Research*, *17*,1340–1350.

Gotschi, E., Delve, R., & Freyer, B. (2009). Participatory photography as a qualitative approach to obtain insights in framer groups. *Field Methods*, *21*(3), 290–308.

Guillemin, M., & Gillam, L. (2004). Ethics, reflexivity, and "ethically important moments" in research. *Qualitative Inquiry*, *10*, 261–280.

Halilovich, H. (2013). Ethical approaches in research with refugees and asylum seekers using participatory action research. In K. Block, E. Riggs, & N. Haslam (Eds.), *Values and vulnerabilities: The ethics of research with refugees and asylum seekers* (pp. 127–151). Toowong QLD: Australian Academic Press.

Harper, D. (2002). Talking about pictures: A case for photo elicitation. *Visual Studies*, *17*(10), 13–26. doi:10.1080/14725860220137345

Hodgetts, D., Radley, A., Chamberlain, K., & Hodgetts, A. (2007). Health inequalities and homelessness: Considering material. *Spatial and Relational Dimensions Journal of Health Psychology*, *12*(5), 709–725.

Hopkins, P. (2008). Ethical issues in research with unaccompanied asylum-seeking children. *Children's Geographies*, *6*(1), 37–48. doi:10.1080/14733280701791884

Immigrant Settlement Association of Nova Scotia (ISANS, 2019). Welcome to ISANS. Retrieved from https://www.isans.ca/about/. Accessed on July 31, 2019.

Johnson, G. A. (2011). A child's right to participation: Photovoice as methodology for documenting the experiences of children living in Kenyan orphanages. *Visual Anthropology Review*, *27*(2), 141–161.

Johnston G. (2016). Champions for social change: Photovoice ethics in practice and "false hopes" for policy and social change. *Global Public Health*, 11(5-6), 799–811.

Jurkowski, J. (2008). Photovoice as participatory action research tool for engaging people with intellectual disabilities in research and program development. *Intellectual and Development Disabilities*, *46*(1), 1–11.

Kabranian-Melkonian, S. (2015). Ethical concerns with refugee research. *Journal of Human Behavior in the Social Environment*, *25*(7), 714–722. doi:10.1080/10911359.2015.1032634

Langmann, S., & Pick, D. (2014). Dignity and ethics in research photography. *International Journal of Social Research Methodology*, *17*(6), 709–721.

Lickiss, N. (2007). On human dignity: Fragments of an exploration. In J. Malpass & N. Lickiss (Eds.), *Perspectives on human dignity: A conversation* (pp. 27–41). London: Springer.

MacKenzie, C., & McDowell, C. (2007). Beyond "do no harm": The challenge of constructing ethical relationships in refugee research. *Journal of Refugee Studies*, *20*(2), 299–319.

Merriam, S. (2009). *Qualitative research: A guide to design and implementation*. San Francisco, CA: Jossey-Bass.

Mulder, C., & Dull, A. (2014). Facilitating self-reflection: The integration of photovoice in graduate social work education. *Social Work Education*, *33*(8), 1017–1036.

Murray, L., & Nash, M. (2017). The challenges of participant photography: A critical reflection on methodology and ethics in two cultural contexts. *Qualitative Health Research 27*(6), 923–937.

Newman, S. D., Maurer, D., Jackson, A., Saxon, M., Jones, R., & Reese, G. (2009). Gathering the evidence: Photovoice as a tool for disability advocacy progress. *Community Health Partnership*, *3*(2), 139–144. doi:10.1353/cpr.0.0074

Noland, C. M. (2006). Auto-photography as research practice: Identity and self-esteem research.*Journal of Research Practice*, *2*(1), 1–19.

Pauwels, L. (2008). Taking and using. *Visual Communication Quarterly*, *15*(4), 243–257.

Peabody, C. G. (2013). Using photovoice as a tool to engage social work students in social justice. *Journal of Teaching in Social Work*, *33*(3), 251–265.

Peck, J. (2016). Vibrant photography: Photographs, actants and political ecology. *Photographies*, *9*(1), 71–89.

Photovoice (2019). Retrieved from https://photovoice.org. Accessed on March 2, 2019.

Povee, K., Bishop, B. J., & Roberts, L. D. (2014). The use of photovoice with people with intellectual disabilities: reflections, challenges and opportunities, *Disability & Society*, *29*(6), 893–907. doi: 10.1080/09687599.2013.874331

Prins, E. (2010). Participatory photography: A tool for empowerment or surveillance. *Action Research*, *8*(4), 426–443.

Radley, A., Hodgetts, D., & Cullens, A. (2005). Visualizing homelessness: A study in photography and estrangement. *Journal of Community & Applied Social Psychology*, *15*, 273–295.

Riessman, C. K. (2005). Exporting ethics: A narrative about narrative research in South India. *Health: An Interdisciplinary Journal for the Social Study of Health, Illness and Medicine*, *9*, 473–490.

Robertson, Z., Gifford, S., McMichael, C., & Correa-Velez, I. (2016). Through their eyes: Seeing experiences of settlement in photographs taken by refugee background youth in Melbourne, Australia. *Visual Studies*, *31*(1), 34–49. doi:10.1080/1472586X.2015.1128845

Singhal, A., Harter, L., Chitnis, K., & Sharma, D. (2007). Participatory photography as theory, method, and praxis: analyzing an entertainment-education project in India. *Critical Arts: South-North Cultural and Media Studies*, *21*(1), 212–227.

Slovic, P., Västfjäll, D., Erlandson, A., & Gregory, R. (2017). Iconic photographs and the ebb and flow of empathic response to humanitarian disasters. PNAS, *114*(4), 640–644.

Smith, V. J. (2009). Ethical and effective ethnographic research methods: A case study with Afghan refugees in California. *Journal of Empirical Research on Human Research Ethics*, *4*(3), 59–72. doi:10.1525/jer.2009.4.3.59

Sontag, S. (1977). *On photography*. New York, NY: Anchor Books.

Strack, R. Magill, C., & McDonagh, K. (2004). Engaging youth through photovoice. *Health Promotion Practice*, *5*(1), 49–58.

Strawn, C., & Monama, G. (2012). Making Soweto stories: Photovoice meets the new literacy studies. *International Journal of Lifelong Education*, *31*(5), 535–553.

Wang, C., & Burris, M. A. (1994). Empowerment through photo novella: Portraits of participation. *Health Education & Behavior*, *21*(2), 171–186.

Wang, C., & Burris, M. A. (1997). Photovoice: Concept, methodology, and use of participatory needs assessment. *Health Education & Behaviour*, *24*(3), 369–387.

Wang, C., Cash, J. L., & Powers, L. S. (2000). Who knows the streets as well as the homeless? Promoting personal and community action through Photovoice. *Health Promotion Practice*, *1*(1), 81–89.

Wiersma, E. (2011). Using photovoice with people with early stage Alzheimer's disease: A discussion methodology. *Dementia*, *10*(2), 203–216.

Wiles, R., Coffey, A., Robinson, J., & Heath, S. (2012). Anonymisation and visual images: Issues of respect, "voice" and protection. *International Journal of Social Research Methodology*, *15*(1), 41–53.

Youth Art Collective (YAC). (2019). About us. Retrieved from https://www.youthartconnection.ca/about-us. Accessed on July 31, 2019.

Zhu, Y. (2019). Can participatory action research empower participants in adult education studies? *The Canadian Journal for the Study of Adult Education*, *31*(1), 63–73.

CHAPTER 9

LETTING IMAGES 'SPEAK FOR THEMSELVES'? USER INTERPRETATION AND THE ETHICAL CONSIDERATIONS FOR SHARING VISUAL DATA ONLINE

Alice Neeson

ABSTRACT

Participatory research approaches often involve visual methods. However, when it comes to data analysis, the opportunities and challenges of participatory visual methods remain underexplored. A focus on the process of data collection, rather than the data itself, may lead to the omission of a formal analysis stage. In action-orientated research, analysis is often discussion-based and undocumented. As a result, the intended 'meaning' behind visual data may not be explicitly communicated to audiences. This chapter explores the ethical considerations of sharing raw visual data. It considers the significance of inter-textual meanings in the digital age, where pathways to accessing that data are highly individualised, and consequently constructed meanings may radically diverge from intended meanings. It also suggests how participatory research facilitators can minimise radical divergence between intention and interpretation.

Keywords: Participatory action research; photovoice; participatory filmmaking; digital media; internet ethics; data analysis

Ethics and Integrity in Visual Research Methods:
Advances in Research Ethics and Integrity, Volume 5, 171–185

ISSN: 2398-6018/doi:10.1108/S2398-601820200000005015

In January 2019, a video of white teenagers wearing 'Make America Great Again' caps surrounding, and apparently mocking, a Native American activist went viral. The footage shows activist Nathan Phillips beating a drum and singing the American Indian Movement Anthem, while a teen[1] in a MAGA hat stands rooted inches away, smiling. The clip was viewed over 2.5 million times before the Twitter account of the original poster was removed. The teenager's behaviour was widely condemned as racist, entitled and disrespectful.

Later, new videos emerged showing that before this encounter, African-American protestors identifying as Black Hebrew Israelites had shouted slurs and taunted the teenagers – students at Covington Catholic High School. Another video shows Phillips slowly approaching the crowd of students, beating his drum. Phillips and the teen in the MAGA hat both explained their version of events: Phillips was aiming to diffuse the tense situation between the young people and the Black Hebrew Israelites, and he felt threatened by the teenagers' behaviour; the teenager who smiled and stared at Phillips said that he purposefully smiled and remained motionless to diffuse the tense situation, while other students were chanting their school spirit chants to drown out the hateful insults of the Black Hebrew Israelites.

Within days of the video first appearing on Twitter, a multitude of viewpoints were being aired on international media: among other claims, the video represented racist entitlement characteristic of 'Trump's America'; it showed the 'liberal media's' demonising of conservative white men; or it revealed the speed at which social media commentators are willing to jump to conclusions in an era of 'fake news'. Despite the emergence of additional footage and the eyewitness testimony that painted a picture of a prolonged, complicated series of interactions, many commentators still interpreted the smile of teen in oppositional manners: it was the smirk of an entitled racist, or it was the bemused expression of a confused kid who didn't know how to react to this bizarre spectacle.

The reactions to the Covington protest video illustrate the risks and challenges for researchers working with visual data – and in particular, researchers who post visual data[2] online. The possibility of oppositional readings of visual data poses a number of important questions for anyone sharing research findings in a visual form: how does context (online and offline) alter how visual data are interpreted and narrativised? To what extent does captioning an image help to 'anchor' the viewer's interpretation? How does the embedding and re-embedding of the image impact upon meaning-making? And, to what extent does the viewer's life-experiences, political viewpoints and social environment stimulate the viewer to construct their own context for the image?

Perhaps more fundamentally, visual researchers should question their responsibility for the after-life of the research project in an era where digital affordances make it simple to for images to be commented on, de-contextualised and re-contextualised. In this context, the ethical implications for visual research do not end with publication and dissemination, but continue as long as outputs can be accessed.

VISUAL DATA IN ACTION-FOCUSSED RESEARCH AND PUBLISHING ONLINE

Visual research methods hold great potential for participatory action approaches, in part due to their emphasis on participant-as-expert and the flexible nature through which participants can express perceptions, opinions, and lived experience. Methods such as photovoice and participatory filmmaking are often employed with the aims of expressing the needs, issues, and desires of people and communities, particularly when these people and communities are socially excluded. The processes of participatory visual research emphasise collaboration, inclusion, and equity in the researcher-participant relationship, making them well suited to research projects relating to community development or conflict transformation.

In recent decades, the internet has provided new opportunities for sharing the outputs of such research projects. This can present a number of advantages: one being that research outputs, which may otherwise have been inaccessible to participants, can be viewed easily and – usually – for free, in contrast to the often high costs of accessing academic publications. This helps to ensure inclusion in the potential learnings and uses of the research project. In addition, publishing online can increase the potential audience for research outputs; a particularly useful opportunity for those projects which seek to amplify the experiences and opinions of socially excluded groups. Although participatory methodologies often place more importance on the process of the research than the outputs, outputs can contribute to the project aims as tools of advocacy and awareness raising. For example, Caroline Wang (2008) writes that as well as offering a process to record a community's assets and challenges, photovoice can 'promote critical dialogue and knowledge about community issues' (p. 147), and can be a means through which to reach decisionmakers in order to effect social change. Likewise, Newman et al. (2009), in their study of photovoice as a tool for disability advocacy, explain:

> Often people with disabilities encounter difficulties in helping the able-bodied population to understand the issues that are faced by persons who use wheelchairs to move about in the community. Our photographs provide visual evidence that helps us to identify and address issues of accessibility in our community .... We are using our photographs to educate the public, as well as those who have the power to make change in our communities. (p. 139)

For these researchers, 'a picture is worth a thousand words' (Newman et al., 2009, p. 139): images, Newman et al. argue, are likely to have a stronger emotional impact on viewers, which in turn makes it more likely that these viewers will support, or potentially become involved with, action for change.[3] Following this argument, the wider the audience, the greater the potential impact.

Visual research methods lend themselves well to action-focussed approaches for a number of additional reasons – one being that they can facilitate the expression of opinions and viewpoints that could be difficult to articulate in words. This can make them particularly useful when working with participants who are communicating in a second or third language (see e.g. Long, Hylton, Lewis, Ratna, & Spracklen, 2011), but also minimises reliance on participants' words as 'magic windows' into their thoughts and behaviours. As Varis (2014) states: 'people do not have an opinion on or a straightforward explanation for everything they do, nor is every aspect of our behaviour easily verbalised' (p. 14). Images, then, present a flexible option for communication which does not rely on education or language skills: participants can show rather than tell what is important to them.

A further advantage of using visual methods for action research projects is that data collation and analysis are generally simpler than with other qualitative methods: participants usually don't have to record and transcribe large quantities of textual data.[4] Indeed, in many cases, both participants and data are allowed to 'speak for themselves' (Aldridge, 2017, p. 28). In other words, visual data are presented in its raw state, without analysis, and sometimes – particularly with participatory filmmaking – without commentary either. This is because the aims of the project are often achieved through the research process. For example, Clemencia Rodriguez (2001) describes how a video produced by striking female workers at a Colombian hospital led to 'shifting power roles' which facilitated a 'collective dynamic that ... challenge[d] institutionalised leadership roles' (pp. 123–124). Similarly, in their 2002 study of Australian community broadcasting, Forde, Foxwell, and Meadows (2002) emphasise how the process of producing content, rather than the content itself, leads to community organisation and political empowerment. They propose that participatory media is a:

> [...] *process* of cultural empowerment ... [where] content production is not necessarily the prime purpose and what may be as (or more) important are the ways in which community media outlets facilitate the process of community organisation. (Forde et al., 2002, p. 3, emphasis in original)

These examples illustrate how participating in media making can be a form of political or civic self-transformation, through which participants become politically aware, politically educated, and politically capable,[5] and through which community formation and community organisation can occur (Atton, 2001, p. 11). Similarly, publishing visual data online can serve a number of benefits for action-orientated research. Siddiquee and Kagan (2006) describe how digital media serves as a 'lifeline' for marginalised groups by countering the 'isolating affects' linked to their treatment by the dominant society (p. 196). Digital media may help to disrupt 'information monopolies' (Radin, 2006, p. 592) through the sharing of alternative information and resources, and through the establishment of counter narratives built from this information. In addition, Balkin (2004) argues that digital technologies can transform public discourse by offering individuals from marginalised groups 'a fair opportunity to participate in the forms of meaning-making that constitute them' (p. 3). For members of socially excluded groups who regularly experience discrimination, digital technology can therefore offer a space where there is potential to feel accepted and understood. This dynamic can bring together connections of people in a way that encourages political engagement and civic interaction (Bentivegna, 2006; Dahlgren, 2005; Dahlberg & Siapera, 2007).

If a research project takes these as goals, sharing visual data on the internet could represent an opportunity for furthering them. While much of the literature on online activism has conceived of the internet as a tool for conventional activism – such as spreading petitions and organising direct action, letter writing campaigns and boycotts (Shaw, 2011; Vegh, 2013), an increasing body of literature is focussing on digital media's potential for discursive activism: that is, activism whereby members of oppressed groups challenge hegemonic narratives, reclaim the right to represent themselves, and build communities based on shared experience and solidarity (e.g. Clark, 2016; Mehra, Merkel, & Bishop, 2004; Shaw, 2011). While this kind of activism may be open to the charge of being ineffective in achieving tangible goals and political outcomes, it may be the case that the impact of discursive activism is simply more difficult to measure than, for example, the size of a crowd at a rally or the amount of media coverage for a cause. Indeed, activists using conventional tactics have recognised the importance of fostering a group identity and critical consciousness (e.g. Levine, 2011; Smucker, 2017).

The emphasis on process in participatory research and the potential effectiveness of sharing visual data in its raw state are likely to form part of the explanation for why participatory forms of data analysis are underdeveloped.[6] Given that analysis of qualitative data is generally laborious and requires some degree of training, there are also practical reasons why researchers may diverge from participatory principles at this stage of the research project. Some may opt for 'traditional', non-participatory approaches to analysis. In these cases, where the data are processed and explained by the 'expert', participation is generally limited to data collection rather than control over stories and final representation. This may not sit easily with the central paradigms of participatory research. As Sreberny-Mohammadi and Mohammadi (1994, p. 228) argue, the manner in which information flows in society is intrinsically linked to power and the exercise of power, and it is therefore impossible to separate the issue of participation in interpretation/communication processes from participation in social and political action. As participatory action research is often conducted with socially excluded individuals and groups, non-participatory analysis misses the opportunity for extending the voices of individuals and groups who are often ignored or alienated by mainstream media (Sreberny-Mohammadi & Mohammadi, 1994, p. 221).

Given the critical attention to voice and representation at the centre of participatory research, a more popular option for researchers may be to present images without analysis, or to separate the images from the analysis. This standalone presentation may be seen as preferable to framing images through non-participatory analysis. For example, in 2016, I coordinated a photovoice project relating to conflict and reconciliation. The resulting images were showcased though an exhibition event in the local community, and as a gallery on the social media profile of a national non-profit organisation. The analysis of the images and the findings of the research, which were initially drafted by a researcher and then edited based on discussions and feedback sessions with participants, were published in a pamphlet. This type of feedback exercise allows for participants to be consulted on research outputs, but as decisions are ultimately controlled by the researcher, it falls short of meaningful participation.

Perhaps unsurprisingly, the pamphlet had a much lower number of readers than the images had viewers. In addition, in many cases, viewer comments on the social media gallery seemed to suggest significantly different interpretations of images to those given by the photographers in the pamphlet. While the photographers spoke of resilience and healing, the online commentators focussed on sorrow, futility, and anger. This led me to question whether, in the researcher's efforts to avoid distorting or misrepresenting the images by facilitating a space for images and participants to 'speak for themselves', they were in fact shifting the narrativising role to the viewers. And viewers, particularly those who engaged with the images online rather than through the community event, were unlikely to have detailed knowledge of the context in which the images were created, and the photographers' intentions in creating them. The different interpretive contexts, in this case, raised doubts as to the utility of publishing the images online. Although participants were pleased with the awareness-raising achieved through social media, some raised concerns that online viewers were using the images to reinforce well-intentioned but stereotypical views.

THE INTERPRETATION OF IMAGES ONLINE

Of course, critical attention to audience interpretation of media did not begin with the internet. Communication theories of an 'active audience', which centre upon the idea of the audience or listener playing an active role in decoding or constructing the meaning of media messages, became prominent in the late twentieth century (see e.g. Hall, 1993; Morley, 1980, 1993). This was a move away from traditional approaches to media and communication theory, which placed emphasis on the production of media texts and their content (Livingstone, 1998). The development of 'new audience research' helped to shift some of this attention from meanings contained within texts, to meanings generated outside of the text. New audience research is primarily associated with qualitative methods, reflecting its focus on the social and cultural contexts in which media is used. As Jensen (1993) states:

> Reception analysis offers insights into the interpretive processes and everyday contexts of media use, where audiences rearticulate and enact the meanings of mass communication. (p. 26)

The internet as a platform offers opportunities for different interpretive processes and everyday contexts to other forms of media. While digital media does not exist within a separate 'virtual world' – rather, it is embedded and imbued with many of the structures and realities of the 'real world' – it does offer certain affordances[7] that raise both possibilities and challenges for visual researchers.

One of these affordances is that digital media blurs the boundaries between media production and media reception, offering communicative possibilities for people to engage with content and other users in a way which was not possible with traditional mass media. This has led some commentators to argue that the concept of 'audience' is no longer useful or relevant for media studies (e.g. Gillmor, 2006; Jermyn & Holmes 2006; McGuinness, 2016). However, following

Livingstone and Das (2013), I argue that although the nature of the 'audience' may have changed, existing approaches for understanding how people use, understand and engage with media remain relevant when applied to digital material.

Livingstone and Das (2013, p. 5) propose that while the new media environment means it is no longer relevant to ask whether people are part of an audience, a focus on individual acts and states of 'audiencehood' can be fruitful. They argue (Livingstone & Das, 2013, p. 4) that audience reception studies bring three significant insights to new media studies. The first is that audience readings of a media text cannot be understood from knowledge of the text alone. This insight reflects the new audience research approach developed in the 1980s and 1990s (along with the associated 'ethnographic turn'), which moved away from media-centric understandings of communication and towards an exploration of local sites of media use and meaning-making. The second insight highlighted by Livingstone and Das (2013) is that interpretation of media texts is shaped by social and cultural contexts, which undermines 'claims of media imperialism and dominant ideologies' (p. 4). This approach emphasises an understanding of media texts as polysemic, an attention to interpretive contexts, and the understanding of audiences as heterogeneous, critical, and active in their interpretations of media texts. The third insight is that the contextualisation of media reception in everyday life allows insight into the 'microtactics of appropriation' that reshape meanings and interpretations. As Morley (1993) argued, 'local meanings are so often made within and against the symbolic resources provided by global media networks' (p. 17).

In addition, Livingstone's (2003) argument that reception and use are two activities which can be examined and understood in a connected manner is particularly relevant for digital media, where use and reception are often blurred as 'users' navigate and select media through multiple pathways. Livingstone (2003) argues:

> [...] people are always both *interpreters* of the media-as-text and *users* of the media-as-object, and the activities associated with these symbolic and material uses of media are mutually defining. And clearly, ethnographic studies of audiences draw on the same insights as reception studies – the stress on active audiences making contingent and context-dependent choices, on fragmentation and plurality within the population rather than merely consumers of the meanings of media – in order to develop the study of the local, typically domestic contexts within which media-as-objects are appropriated as part and parcel of everyday life. (p. 343)

One of the most striking examples of 'use' of online media is in the appropriation of digital material to further political arguments. The Covington protest video, for example, has been shared on Twitter with criticisms of 'the youth today', 'racist Tump supporters', 'biased media', and 'left wing mobs'. As evidenced by the heated debates around the footage, the meaning of the video is not solely contained within in, but constructed around it based on competing narratives and worldviews.

Furthermore, online content can be created or appropriated to present not only different interpretations of events, but also different factual realities. Interested in the popular media discourses about 'fake news', I conducted a small study with social media users who shared a story that contained false information and false allegations. Much of the public debate on fake news has assumed that the social media users who read and share the stories are unable to differentiate between what is real and what is fake.[8] However, of 15 people that I interviewed, 13 believed or suspected that the story they had shared was not true. Their motivation for sharing the story was not that they thought it contained accurate or little-known information, but because the story fit with their political viewpoint, and they believed that sharing it may convince others of the arguments contained within it. This poses a question about how digital media users actively and strategically choose to share content that furthers personal and political aims; a question that could be worth pursuing for researchers interested in the circulation of fake news.

The affordances of digital media, which make it possible for users to actively create or appropriate content, help to shape an interpretive environment which may be difficult to predict before posting. With a physical display of visual data (such as an exhibition), a researcher may be able to predict and, to a large degree control, how widely an image is shared, who views it, and how likely it remains viewable. However, with a digital display, predicting or controlling these elements becomes very difficult.

THE INTERPRETIVE CHALLENGE FOR ACTION RESEARCH

The structure of digital media is dispersed rather than centralised, and hence the content and flow of communication cannot be easily controlled. Whereas the mass media is generally produced only by professional journalists, digital media production is, in theory, possible for almost anyone with an internet connection.[9] Following from this, social groupings online tend to be egocentric rather than structurally distinct – that is, social networks are structured with the individual at the centre of their own community, and follow a trajectory according to the individual's actions and behaviours (boyd & Ellison 2007, p. 9). In most cases, online spaces are not structured as bounded communities, and instead users navigate and connect with others according to their needs and interests. Postill and Pink (2012) maintain that digital media is therefore best understood in terms of sociality rather than community – that is, in terms of social relations experienced as emplaced and socially contextualised interactions rather than as membership of a bound cluster. On some platforms, the uniqueness of these interactions and pathways is further reinforced by algorithms, which give prominence to content based on the assumed interest of the user (see e.g. Flaxman, Goel, & Rao, 2016; Zuiderveen Borgesius et al., 2016).

Social media makes it easy for users to share visual data with their own comments, which can encourage certain 'readings' and interpretations. Of course, in printed media, image and text also interact with each other to encourage particular readings. However, the diffusion and quantity of digital media results in more dispersed 'interpretive communities' (Lindlof, 1988), meaning there are fewer shared frameworks and forms of discourse for interpreting media. Image editing tools also make it easy for users to change or manipulate data in a way that encourages diversion between intention of the media creator and interpretation by viewers.

Outside of social media platforms, hyperlinking is another way that digital media content is connected and placed in a context of wider social, economic, and political issues. Hyperlinks are words or images that, when clicked, navigate a user to a different digital location. Hypertext, as Snyder (1998) explains, is:

> [...] a structure composed of blocks of text connected by electronic links, it offers different paths to different users ... the extent of hypertext is unknowable because it lacks clear boundaries and is often multi-authored. (pp. 126–127)

This allows users to consume content in unique paths and patterns. Navigating through self-selected spaces online, users are able to construct autonomously their own viewing and media consumption experiences. Livingstone and Das' (2013) analysis of contextualisation, polysemy, and appropriation would suggest that these affordances lend themselves to multiple interpretations, some of which may be oppositional to the media creator's intended meaning. The inter-textual and extra-textual meanings intended by media creators can easily be lost or distorted in the highly individualised contexts generated online. This is an important ethical consideration for visual researchers: appropriation of visual data could lead to stereotyping, online abuse, or misconceptions.

In addition, the loss of control over content by the publisher can make it difficult for digital media to serve as a vehicle for social change, thus undermining some of the potential of action research projects. Traditional mass information or persuasion campaigns aim to promote collective goals for social change, but this is complicated by the egocentric nature of the internet, where individuals can self-select content that corresponds to their worldview. In these circumstances, support for causes may be conditional on individual interest. Collective efforts for change require a degree of connectedness that does not always follow from individual reflection or critical listening, as useful as these activities may be for conscientisation. Ewick and Silbey (1995) argue that linking individual stories has subversive potential by uncovering the relationship between individuals, social organisation, and structural inequality. It follows that when media is atomised or arranged in ways different to those intended by media creators, it can lose this subversive potential. On the other hand, digital media affordances may also serve to facilitate social and narrative assemblage – for example, by codifying separate pieces of content through hashtags. The development of collective goals through online platforms, therefore, remains possible but unpredictable.

SUGGESTIONS FOR CONDUCTING PARTICIPATORY VISUAL RESEARCH AND FOR FUTURE WORK

The challenges for researchers working with participatory visual media are multiple, but there are also numerous possible rewards. In addition to rendering visual research more accessible, sharing data online can facilitate advocacy, awareness raising and – potentially – action for social change. Below are a number of suggestions for consideration by researchers working with participatory visual data.

Further Research and Experimentation in Participatory Techniques for Data Analysis

Experimentation and innovation in participatory analysis methods is required in order to fully achieve the disruptive potential of participatory research. The design and implementation of participatory visual research has been widely explored, but a limited number of studies have focussed on participatory approaches to visual data analysis. Without appropriate attention to this step of the research process, narrativising of data can be misappropriated or deflected. Potential avenues for exploration might include online interfaces that facilitate interactions and reduce the amount of time required for analysis; and live data visualisation which makes analysis more engaging and easier to understand.

Meaningful Involvement of Participants in Communication and Dissemination of Research

In addition to facilitating participation in the analysis of research data, participatory approaches should also incorporate participation in the communication and dissemination of research findings and data. Visual data have powerful potential for advocacy and awareness raising, but without the meaningful involvement of participants, this potential can be misplaced or lost. Particularly when working with images that are potentially sensitive, the researcher should ensure that participants have adequate time to reflect and, if desired, receive advice on the potential reception before giving consent to disseminate. Researchers should be prepared that participants may not wish to publish or disseminate the research beyond the community. If possible, this consideration should be made at proposal or bid stage, to avoid imposing requirements that the participating community is not comfortable with. Researchers may also wish to explore alternative outputs with participants. For example, a participant may not wish their image to be disseminated, but may be keen to promote an action that resulted from visual research.

Careful Consideration about the Use of Online Platforms for Publication of Images

When deciding if and where to communicate research outputs, particular consideration should be given as to whether the internet is the best platform to share visual data. As this chapter has highlighted, there are certainly benefits to publishing data online. However, certain affordances of online media also carry extra risks, and it is rarely possible to predict how images might be interpreted, where they might be shared, or how they might be appropriated. This is a particularly important consideration when research relates to a politically contested issue, or involves a socially excluded community. Where possible, this decision should be made in an informed manner by research participants, with guidance and advice from researchers.

Provision of Advice to Participants about Expected Levels of Engagement

Where possible, researchers should provide advice on how widely an image is likely to be shared, how long the image will be made available, and who the potential audience is likely to be. Although the nature of the internet makes these considerations difficult to predict, efforts should be made to produce informed estimates based on past and existing analytics (such as Google Analytics, Facebook Audience Insights, Twitter Analytics, and Instagram Insights). Participants should also be clearly advised about the levels of unpredictability when sharing content online.

Monitoring and Tracking of Images Published Online

The researcher may also wish to consider whether they are prepared to track (or provide training and resources for participants to track) if and how images are shared online, using tools such as Pixsy or Tineye. Notice and take down processes could be initiated in cases of libel or copyright infringement. This would not, however, prevent internet users from adding comments with hyperlinks to the original image. It should be made clear to participants that although researchers should be able to remove the original image if consent to share is withdrawn, copies and separately uploaded versions may be more difficult to remove. This does not mean, however, that efforts cannot be made to trace images and request their removal.

Inclusion of Editorial Guidance with Images

When sharing visual data online, consideration should be given to the inclusion of editorial guidance. For example, still images could be embedded within explanatory videos, or participatory films could be overlaid with explanatory text created by the participant, with support from the researcher if required. While this does not guarantee that images will be viewed together with their captions or context (screen shots and simple video clipping tools make it easy to capture and remove individual elements), it may help to minimise interpretations of images that are radically divorced from the meanings that participants intended to convey.

CONCLUSION

In the early twentieth century, Soviet filmmaker Lev Kuleshov carried out a series of informal experiments with film montage. In one, he and his team edited a series of clips to show an identical shot of an actor's expressionless face juxtaposed with various objects and subjects: a bowl of soup, a child playing with a toy, and a woman in a coffin. Viewers interpreted the same clips of the actor in different ways depending on the order in which it was shown: he was hungry when edited next to the bowl of soup, happy when edited next to the child, sad when edited next to the woman in the coffin. Kuleshov claimed that his explorations demonstrated the power of editing.

Since the popularisation of the internet, media producers have less control over the context into which images are edited, which is significant because the meaning of visual data can exist extra-textually and inter-textually and in ways that are often beyond the control of the researcher. On one level, reception theory has highlighted how audience interpretations are influenced by their social and cultural contexts, and that 'micro-tactics of appropriation' (Livingstone & Das, 2013, p. 4) can be used to construct local meanings which may differ or sit in opposition to intended meanings. This tendency is not unique to online environments, although digital media can contribute to the shaping of social and cultural contexts in radical ways.[10]

On another level, the internet allows users to construct their own engagement pathways, perhaps clicking on hyperlinks from commentaries or clicking on a link that has been attributed a hashtag suggesting a particular perspective. When accessing visual data in these digital contexts, the media may be interpreted in different ways to those intended by its creators. This can create some tension between digital media practice and the development of collective goals for change, so is a particular concern for researchers undertaking action research. In addition, the ease for users to 'add their own voice' to an online conversation, perhaps by posting a comment on a website, or sharing a link on social media with a comment, creates a risk for researchers who aim to allow the data (or participant's story) to 'speak for itself'.

NOTES

1. I have not used the young person's name here because he was under 18 at the time of the event, and the Convention on the Rights of the Child (1989) enshrines a child's right to privacy. However his identity was widely reported on and, perhaps as a result, his family hired a public relations firm and he participated in a media interview. The potential for children to be 'named and shamed' online raises an ethical consideration for the publishing and sharing of images of children, particularly when that child is at risk of stigmatisation. Not naming a child is often an inadequate measure to protect their identity, and before posting an image, adults should consider how widely it might be shared as well as the potential impacts – both short and long-term – on the child.
2. Images produced as part of the research process could also be classed as research outputs rather than data. In using the term data, I recognise the utility of 'data as output', and disagree that there is necessarily a clear distinction between the two categories. By visual data, I refer to still and moving images which tell a story, but do not include an analysis of that story – which contrasts to traditional qualitative research outputs, which generally include both story and analysis.
3. Of course, engaging with an issue online and taking action on the issue is not often a rapid, sequential pathway. Popular and academic attention to 'slacktivism', 'clicktivism' and 'virtue signalling' (see e.g. Budish, 2012, Christensen, 2011, Kristofferson, White, & Peloza 2014) focus on the complexities of activism in the internet age, and the differences between conscientization and action.
4. This said, the traditional process of recording, transcribing, and coding is not the only possible route for analysing textual data. For example, a grouping activity can help to identify patterns and themes in pre-transcribed snippets of data, or a ranking activity can help to identify priorities and values, as well as levels of resonance with a group or community. Producing these snippets does usually require a first level of analysis in order

to summarise or synthesise data – an activity which can be time consuming and may be difficult to optimise for participation.

5. 'Political' here is used in the Gramscian sense: it refers not to merely holding a view about a particular situation, but to seeking to alter the terrain of power (see Gramsci, 1991): a commonly held goal of participatory action approaches.

6. Some participatory visual methods do incorporate forms of analysis as part of their central process. For example, photovoice includes a summative discussion and captioning exercise with participants. Participatory Video and the Most Significant Change (Asadullah & Muñiz, 2015) (a participatory evaluation technique in which a group produces films and reflects on stories of significant change) includes collective sorting, categorising, and prioritising. However, most of this analysis is discussion-based rather than written, meaning that future viewers of the data may not have access to the analysis that was conducted. Analysis, in this sense, takes the form of reflecting on learnings in order to chart subsequent actions; it is a means to an end rather than for the purpose of an output.

7. Following Nagy and Neff (2015), I use the term 'affordance' as a socio-technological concept that refers to both the material possibilities offered by technology, and the active participation of humans interacting with them.

8. See, for example, the *BBC*'s article 'Schools should teach pupils how to spot fake news' (Coughlan, 2017); *USA Today*'s article 'How to spot fake news' (2017); *Forbes*' article 'How and why brands need to differentiate between real vs. fake news' (Olenski, 2017); TED's blog post 'How to tell real news from fake news' (McClure, 2017); *Huffington Post*'s article 'Fake news is real. Here's how to know if you're reading it' (Jackson, 2017); *The Guardian*'s quiz 'Can you spot the 'real' fake news story?' (Haynes, 2016); or the *Financial Times*' quiz 'Fake news or not: Can you tell what is real?' (Tassell, Blood, & Wilson, 2016).

9. This is not to claim equality of access and participation in digital media. The dominant direction of communication flows from media conglomerates, for example, rather than 'user-generated' content. However the internet is not simply a democratic public sphere, and structural power inequalities that exist in wider society are frequently replicated online. A great imbalance of communications capacity exists. This conforms to traditional media patterns, such as that of large audiences visiting a small number of sites.

10. For example, it is difficult to imagine that such a rapid recognition of the extent of sexual abuse and harassment could take place without the affordances of social media, which allowed women from all walks of life to share their experiences, and connect these experiences under the #MeToo hashtag, showing their prevalence and the systemic nature of the issue.

REFERENCES

Aldridge, J. (2017). Advancing participatory research. *Relational Social Work*, *1*(2), 26–35.

Asadullah, S., & Muñiz, S. (2015). *Participatory video and the most significant change*. Oxford: InsightShare.

Atton, C. (2001). *Alternative media*. London: Sage.

Balkin, J. (2004). *Digital speech and democratic culture: A theory of freedom of expression for the information society*. Yale Faculty Scholarship Series 240. Retrieved from http://digitalcommons.law.yale.edu/fss_papers/240/

Bentivegna, S. (2006). Rethinking politics in the world of ICTs. *European Journal of Communication, 21*, 331–343.

boyd, d., & Ellison, N. (2007). Social network sites: Definition, history and scholarship. *Journal of Computer-Mediated Communication*, *13*(1), 210–230.

Budish, R. H. (2012). Click to change: Optimism despite online activism's unmet expectations. *Emory International Law Review*, *26*(2), 745–772.

Christensen, H. S. (2011). Political activities on the internet: Slactivism or political participation by other means? *First Monday*, *16*(2), 1–10.

Clark, R. (2016). "Hope in a hashtag": The discursive activism of #WhyIStayed. *Feminist Media Studies*, *16*(5), 788–804.

Convention on the Rights of the Child (CRC). (1989). Resolution 44/25, opened for signature 20 November 1989, entered into force 2 September 1990.

Coughlan, S. (2017). Schools should teach pupils how to spot fake news. BBC News. Retrieved from http://www.bbc.co.uk/news/education-39272841

Dahlberg, L., & Siapera, E. (Eds.). (2007). *Radical democracy and the internet: Interrogating theory and practice*. London: Palgrave Macmillan.

Dahlgren, P. (2005). The internet, public spheres and political communication: Dispersion and deliberation. *Political Communication*, *22*, 147–162.

Ewick, P., & Silbey, S. (1995). Subversive stories and hegemonic tales: Toward a sociology of narrative. *Law and Society Review*, *29*(2), 197–226.

Flaxman, S., Goel, S., & Rao, J. (2016). Filter bubbles, echo chambers, and online news consumption. *Public Opinion Quarterly*, *80*(1), 298–320.

Forde, S., Foxwell, K., & Meadows, M. (2002). Creating a community public sphere: community radio as a cultural resource. *Media International Australia*, *103*, 56–67.

Gillmor, D. (2006). *We the media: Grassroots journalism by the people, for the people*. Sebastopol: O'Reilly Media.

Gramsci, A. (1991). *Prison notebooks*. New York, NY: Columbia University Press.

Hall, S. (1993). Encoding, decoding. In S. During (Ed.), *The cultural studies reader* (pp. 90–103). London:Routledge.

Haynes, G. (2016). Can you spot the "real" fake news story? The Guardian. Retrieved from https://www.theguardian.com/theguardian/2016/dec/28/can-you-spot-the-real-fake-news-story-quiz

How to spot fake news. (2017). *USA Today*. Retrieved from http://www.usatoday.com/story/news/2017/02/13/fake-news-how-to-tell/97846900/

Jackson, N. (2017). Fake news is real. Here's how to know if you're reading it. *Huffington Post*. Retrieved from https://www.huffingtonpost.co.uk/entry/how-to-find-fake-news_us_589223d6e4b0e35f0fb3c821?ri18n=true&guccounter=1

Jensen, K. (1993). The past in the future: Problems and potentials of historical reception studies. *Journal of Communication*, *43*(4), 20–28.

Jermyn, D., & Holmes, S. (2006). The audience is dead; long live the audience! Interactivity, "telephilia" and the contemporary television audience. *Critical Studies in Television*, *1*(1), 49–57.

Kristofferson, K., White, K., & Peloza, J. (2014). The nature of slactivism: How the social observability of an initial act of token support affects subsequent prosocial action. *Journal of Consumer Research, 40*(6), 1149–1166.

Levine, B. (2011). *Get up, stand up: Uniting populists, energising the defeated, and battling the corporate elite*. Hartford, Vermont: Chelsea Green Publishing.

Lindlof, J. R. (1988). Media audiences as interpretive communities. In J. Anderson (Ed.) *Communication yearbook* (Vol. 11, pp. 81–107). Newbury Park, CA: Sage.

Livingstone, S. (1998). Audience research at the crossroads: The "implied audience" in media theory. *European Journal of Cultural Studies*, *1*(2), 193–217.

Livingstone, S. (2003). The changing nature of audiences: From the mass audience to the interactive media user. In A. Valdivia (Ed.) *Companion to media studies* (pp. 337–359). Oxford: Blackwell.

Livingstone, S., & Das, R. (2013). The end of audiences? Theoretical echoes of reception amid the uncertainties of use. Retrieved from http://www.lse.ac.uk/media@lse/WhosWho/AcademicStaff/SoniaLivingstone/pdf/Livingstone-and-Das,-manuscript.pdf

Long, J., Hylton, K., Lewis, H., Ratna, A., & Spracklen, K. (2011). Space for inclusion? The construction of sport and leisure spaces as places for migrant communities. In A. Ratna & B. Lashura (Eds.), *Community and inclusion in leisure research and sport development* (pp. 33–53). Leeds: Leisure Studies Association.

McClure, L. (2017). How to tell real news from fake news. TEDEd. Retrieved from http://blog.ed.ted.com/2017/01/12/how-to-tell-fake-news-from-real-news/

McGuinness, P. (2016). The people formerly known as the audience: Power shifts in the digital age. *Communications Research and Practice*, *2*(4), 520–527.

Mehra, B., Merkel, C., & Bishop, A. (2004). The internet for empowerment of minority and marginalised users. *New Media and Society*, *6*(6), 781–802.

Morley, D. (1980). *The "nationwide" audience: Structure and decoding*. London: BFI.
Morley, D. (1993). Active audience theory: Pendulums and pitfalls. *Journal of Communication, 43*(4), 13–19.
Nagy, R., & Neff, G. (2015). Imagined affordance: Reconstructing a keyword for communication theory. *Social Media and Society, 1*(2), 1–9.
Newman, S., Maurer, D., Jackson, A., Saxon, M., Jones, R., & Reese, G. (2009). Gathering the evidence: Photovoice as a tool for disability advocacy. *Progress in Community Health Partnerships: Research, Education, and Action, 3*(2), 139–144.
Olenski, S. (2017). How and why brands need to differentiate between real vs. fake news. Forbes. Retrieved from https://www.forbes.com/sites/steveolenski/2017/01/31/how-and-why-brands-need-to-differentiate-between-real-vs-fake-news/#f2b8edb53e9c
Postill, J., & Pink, S. (2012). Social media ethnography: The digital researcher in a messy web. Retrieved from http://blogs.bournemouth.ac.uk/research/files/2013/04/Postill-Pink-socialmedia-ethnography.pdf
Radin, P. (2006). "To me, it's life": Medical communication, trust, and activism in cyberspace. *Social Science and Medicine, 62*, 591–601.
Rodriguez, C. (2001). *Fissures in the mediascape: An international study of citizens' media*. Cresskill, NJ: Hampton Press.
Shaw, F. (2011). (Dis)locating feminisms: Blog activism as crisis response. *Outskirts: Feminisms along the Edge, 24*. Retrieved from http://www.chloe.uwa.edu.au/outskirts/archive/volume24/shaw
Siddiquee, A., & Kagan, C. (2006). The internet, empowerment, and identity: An exploration of participation by refugee women in a community internet project (CIP) in the United Kingdom (UK). *Journal of Community and Applied Social Psychology, 16*(3), 189–206.
Smucker, J. (2017). *Hegemony how-to: A roadmap for radicals*. Chico, CA: AK Press.
Snyder, I. (1998). Beyond the hype: reassessing hypertext. In I. Snyder (Ed.), *Page to screen: Taking literacy into the electronic era* (pp. 125–143). London: Routledge.
Sreberny-Mohammadi, A., & Mohammadi, A. (1994). *Small media, big revolution: Communication, culture and the Iranian revolution*. Minneapolis, MN: University of Minnesota Press.
Tassell, T., Blood, D., & Wilson, K. (2016). Fake news or not: Can you tell what is real? Financial Times. Retrieved from https://ig.ft.com/sites/quiz/fake-news-or-not/
Varis, P. (2014). Digital ethnography. *Tilburg Papers in Cultural Studies, 104*, 1–21.
Vegh, S. (2003). Classifying forms of online activism: The case of cyberprotests against the World Bank. In M. McCaughey & M. Ayers (Eds.), *Cyberactivism: Online activism in theory and practice* (pp. 71–96). Hove: Psychology Press.
Wang, C. (2018). Youth participation in photovoice as a strategy for community change. *Journal of Community Practice, 14*(1–2), 147–161.
Zuiderveen Borgesius, F., Trilling, D., Moeller, J., Bodó, B., de Vreese, C., & Helberger, N. (2016). Should we worry about filter bubbles? *Internet Policy Review: Journal on Internet Regulation, 5*(1), 1–16.

CHAPTER 10

THE ETHICS OF SCREEN SCHOLARSHIP

Jennie Carlsten

ABSTRACT

Image-based researchers have previously noted the failure of institutional ethics guidelines and frameworks to adequately address the issues and methodologies used in visual research, citing a need for more creative and flexible approaches. Screen scholars work with traditional (non-visual) data sources, human subjects, visual data, and colleagues. This produces a range of ethical issues which are not addressed by institutional frameworks, based as they are on individualism and utilitarian ideals. Drawing on the ideas of Johann Neem and Savannah Dodd, this chapter suggests the need to shift our research practices towards a more collective, reflective, and empathetic mode of working.

Keywords: Screen studies; ethics; academic practice; film scholarship; collegial ethics; situated ethics

As screen scholars and academics, we may begin a project assuming that there are institutional frameworks and guidelines in place to preserve the integrity of our research, or that our own moral compass will be sufficient to carry us through. By adhering to these, surely, we can keep within the ethical boundaries drawn by those who came before us. Before too long, though, we might notice that no one is requiring us to adhere to any particular guidelines; a freedom from oversight which might come as a relief or, equally and perhaps simultaneously, as a concern. Alternatively, we might find that the guidelines with which we are compelled

Ethics and Integrity in Visual Research Methods:
Advances in Research Ethics and Integrity, Volume 5, 187–198

ISSN: 2398-6018/doi:10.1108/S2398-601820200000005016

to comply are insufficient, inapplicable, or inappropriate for the research methodology or questions we wish to pursue, leaving us with a bureaucratic burden but not a corresponding clarity of purpose.

The ethical frameworks put in place for researchers have generally been designed by and for the hard and social sciences. Screen scholarship raises unique ethical concerns and obstacles. Perhaps, even, it offers new possibilities for thinking about ethical issues, and how we, as individuals, can be more ethical researchers. Certainly, many image-based researchers have spoken of the need for more creative or flexible approaches to research ethics.

In this chapter, I will discuss some of those creative or flexible approaches; more are detailed by the other contributors to this volume. Further, I suggest a change in thinking about our own work as scholars; redirecting our research practices away from individualism and utilitarianism, and towards a more collective, reflective, and empathetic mode of working that runs throughout scholarship from bottom to top.

I am not speaking here about (or, at least, not primarily or directly about) image-making and research-by-practice. Others in this volume have done so with authority and insight drawn from their own experience as filmmakers and photographers. Rather, I hope to address visual research ethics from the perspective of another group of researchers, those who study the images created by others in order to uncover information about society and its values, ideal or behaviours, interpreting meaning through narrative and formal qualities of the image-as-text.

Such scholarship necessarily engages with multiple spheres of activity: 'traditional' (non-visual) data sources; human subjects (in terms of both individual living research subjects, and the communities affected by our research practices and outputs); visual data (the film, video and photographic artefacts studied); and the colleagues with whom we share our research and institutional spaces. Each of these spheres demands our attention. If we are to move towards a more collective, engaged, and reflective mode of research, there must be changes not only at the level of the institutions and frameworks in which we operate, but in our own lower-level actions. This means attention to:

- how research is valued and evaluated;
- how ethics approval is managed;
- how we analyse the visual data we're studying;
- how we think about the subjects and the end-users (consumers or spectators) of those images;
- what we expect from our colleagues and ourselves in the workplace; and
- how we share our research.

This shift in our approach to image-based research is, ultimately, a reflection of the changes we see happening in the 'real world' at a time of global and environmental crisis. As we think differently about the world and our place in it, our research practices, too, should change. As screen scholars, we pay lots of attention to the ethics *of* the works we look at – but a lot less to how we ethically interact with them.

Screen scholarship ethics is an underdeveloped area. The thinking about image-based research ethics has been largely led by those who work in a social science mode (for whom the aesthetics and form of the image are secondary to its evidentiary role). It also borrows heavily from the conventions of a scientific model which has been codified by institutional practice and procedures. The professional guidelines that have been developed or adopted, as Wiles et al. explain, provide a touchstone or supplementary starting point for researchers:

> The specific ethical issues that researchers identify in their research are informed by their own moral outlook and their understanding of ethics in research (so they can be understood as ethical issues or moral issues). The frameworks for thinking about and managing them are informed largely by the ethical principles derived from the various approaches to ethics which are set out in professional ethical guidelines as well as various textbooks on the topic. Some of these ethical issues can be considered prior to the research commencing but many are emergent and only become apparent as the research proceeds. Researchers can draw on a range of resources from the literature and the research community to assist their thinking in how to manage such issues. It is crucial that they resolve the issues in ways that accord with their moral beliefs but also in ways that do not contravene the established ethical standards of their profession. (Wiles et al., 2008, p. 6)

In thinking about how we can approach our own scholarship more ethically, it is necessary first to consider the philosophical drives of the institutions in which we work. The institutional approach to research ethics is driven by a utilitarian philosophy, following on Bentham and Mill, which judges actions based upon their consequences. An action is right if it produces the consequence of happiness for the greatest number of people, and the consequence of harm for the smallest number of people. An approach to ethical research derived from utilitarianism evaluates practices based on the costs and benefits to individuals, and seeks to 'ensure harm is minimised or avoided and that benefit is maximised' (Iphofen, 2013, p. 13). This is the prevailing model of ethics, to which academics are expected by their institutions to adhere.

Johann Neem's (2013) essay 'Making Sense of the Higher Education Debate' contrasts utilitarian and pragmatic approaches to higher education with a model based on Aristotelian virtues. 'A utilitarian approach to higher education … assumes that colleges and universities must be primarily outward looking, responding to the wishes of higher education clients' (Neem, 2013). This is the approach fostered by a for-profit education system, and (reluctantly or otherwise) by policymakers and university administrators. Writing at the time of the Obama administration in the United States, Neem describes that administration's approach as an example of the pragmatic approach. A pragmatist approach to universities defines their usefulness and success through their outcomes; specifically, their contribution to employment and economic measures. As Neem notes, this pragmatic approach is not concerned with any internal or inherent values to knowledge and learning.

Screen scholarship, especially, finds itself pulled between competing models. As universities seek to monetise screen scholarship through a greater focus on (and appeal to) industry, and as practice-based research expands to draw in more researchers from outside the traditional humanities base, these tensions increase. Researchers, increasingly, have been trained in and acclimatised to work cultures which emphasise utilitarian or pragmatic values rather than Aristotelian ones. The neoliberal normalisation of casualised labour, meanwhile, forces individuals into increasingly difficult circumstances and isolates researchers from one another – effectively silencing dialogue, undermining individual agency, and removing the physical and discursive spaces for debate or dissent around ethical concerns.

Neem is an advocate of a third philosophical position, that of 'virtue ethics'. Virtue ethics are character-based, rather than derived from the consequences of actions (as in utilitarianism or pragmatism) or from authoritarian restrictions and duties (as in, for instance, the model of the university ethics review committee). Neem points out that, under a virtue ethics framework, the means and the ends are inseparable; 'the means are the practices that sustain the virtues' (Neem, 2013). Applying this to the institution of the university, he goes on to argue that unless it sustains and values 'intellectual virtues', the university is not really, any longer, a university. The defenders of intellectual virtues, says Neem, are the faculty: 'Professors thus sustain ways of thinking and doing that are threatened in a society dominated by utilitarianism and pragmatic modes of thought' (Neem, 2013). While Neem is speaking here of an approach to higher education within the context of the Western democratic university system, I propose that his observations are relevant to the culture of research ethics.

Neem here presents the university researcher and educator as a heroic figure, and one in the Arthurian mode – preserving good by being a good and virtuous individual. Do we need to be Galahad to be an ethical researcher? Or does that standard, as for Sir Lancelot, set us up for a fall?

Clearly, there is room for a more robust approach to image-based research, for practitioners and for those who use other's images in their work. Savannah Dodd's work on ethics and photography suggests three guiding principles: *empathy, autonomy,* and *integrity*, principles which Dodd argues can guide individual choices more effectively than checklists or institutional guidelines. Dodd's Photography Ethics Centre teaches 'ethical literacy' to photographers and researchers, using the three principles to illustrate and help students work through ethical problems to arrive at their own 'ethically literate' conclusions (Dodd, 2019). These qualities of empathy, autonomy, and integrity, it strikes me, are characteristics not of research, but of researchers – are they, perhaps, the sort of ethics values to which Neem refers? By acting as autonomous, empathetic people of integrity, will we necessarily produce research which is itself ethical?

'TRADITIONAL' DATA RESEARCH PRACTICES

Screen scholars naturally make use of non-visual sources: for example, text-based resources (books, journals, and archival documents), industry figures, statistics, and demographic data. The ethical guidance around the use of these materials, relatively speaking, is more robust. Most research institutions have both formal frameworks – sets of research standards and training for academic researchers – and informal frameworks – the social networks and research culture promoted within schools and faculties – that are accessible to the researcher. Many of these derive from and depend upon the place of screen scholarship within the particular institution: is the film department housed within the school of communications, or of creative arts? Are media courses the province of the school of English, or housed alongside journalism and TV production? The shifting and uncertain nature of the discipline (or more properly, perhaps, disciplines) means that there is great variety in the attitudes and resources available to the would-be researcher, even around text-based research. When it comes to ethical guidelines around image-based research material, this is even more problematic. The interdisciplinary nature of screen scholarship adds to the complexity of the issues faced by the researcher, and may also pose a challenge as researchers find themselves working in fields outside their main area of experience and training, and must learn the rules and institutional procedures of multiple disciplines.

Aimed as they are at regulating the interactions between the researcher and human subjects, the existing institutional frameworks emphasise the principle of 'do no harm'. These frameworks rely on a medical and scientific model of research rooted in positivism, which presumes 'the existence of a putative knowable reality, and that objective, universal truths can be revealed through empirical scientific data collection and explicit, transparent, experimental research operations and procedures' (Halse & Honey, 2007, p. 131). These frameworks are designed by funding bodies, universities, research and professional organisations, and central government departments that are responsible for particular participant groups (such as, but not limited to, prisoners, healthcare providers, or members of the armed forces). At the level of the university, there are frequently multiple stages of review required. These might include internal peer review of research proposals; budgetary approval; formal risk assessments; employment procedures; and oversight of any publication of findings (Boddy et al., 2010). The goal of these institutional frameworks is the protection of human subjects, but also the institution itself. Universities and professional organisations, and their ethics committees, are risk-averse, yet research that deals with human emotions and behaviour requires risks on the part of both subjects and researchers.

Andrew Clark has written about the shortfalls of existing frameworks. He identifies three particular challenges posed by image-based research (Clark, 2013). First, visual data differs in form, function and mode of production from word-based or numerical data, and these differences engender their own ethical conflicts. Second, the existing institutional procedures and codes of practice have not been designed with visual data in mind. And third, there is a lack of clarity about what researchers should do when they encounter this mismatch. Clark argues that 'it is at the point where visual methodology and "conventional" or well established ethical practices diverge' that moments of 'ethical anxiety' emerge (ibid., p. 69).

The solution Clark offers is a situated ethics. Rather than hard and fast guidelines, he advocates considering projects on a case-by-case basis, noting 'the importance of contextually appropriate ethical actions and decision-making, recognising research ethics as part of an on-going process of negotiation, reflection, and experimentation' (ibid., p. 76). Clark cautions against relativism, acknowledging that some absolute guidelines are necessary, but he also points out that the ethics of visual images are something we encounter and must consider not only in the research environment, but in the rest of our daily lives, our 'everyday visual practice'.

Clark's proposal for a situated ethics draws on his own experiences as a researcher working with vulnerable human subjects. He explores ideas of informed consent, anonymity, and confidentiality, and notes the inevitability of power dynamics and bias in the researcher/participant relationship. He further points out that the intent of the researcher is insignificant next to the potential (future, unknown, or hypothetical) circumstances, use and affect of the circulated or recirculated images. Under these conditions, methodological ethics are rarely straightforward: Clark points out, for instance, that obtaining informed consent is problematic when neither researcher nor participant can predict the future uses of images with absolute certainty; while the practice of anonymising images may dehumanise or further marginalise the subject. The use of images in or as research is inextricable from power dynamics and interpersonal relationships, and these cannot be accounted for by absolutist approaches.

In her essay '"The Project Cannot Be Approved in Its Current Form": Feminist Visual Research Meets the Human Research Ethics Committee', Penelope Pitt argues that a situated ethics model is not sufficient to deal with the dilemma faced by visual researchers, because it still relies upon patriarchal and overly prescriptive value systems, as well as the positivist biomedical model of research. Pitt, who encountered difficulty when she sought ethics approval for a photography project involving members of her family, rejects the ethic committee's ideals of professional detachment and impartiality:

> The advisor suggested that I would find it difficult to remain impartial, and that this may be a threat to the validity of my findings. As I had no intention of taking on the role of impartial observer as researcher, I dismissed this concern. It seemed to me that the advisor had either missed the sections where I described my study as a feminist ethnography or was unfamiliar with this epistemological viewpoint and methodology. (2014, p. 316)

Pitt places her research within the context of a feminist ethics that incorporates the human relationship between researcher and subject, and involves the subjects in the development of the project and approach. In doing so, she draws on work by Christine Halse and Anne Honey.

Halse and Honey, in their article 'Rethinking Ethics Review as Institutional Discourse', argue convincingly that the institutional model of research ethics is inherently and irretrievably flawed; unable to provide effective and appropriate guidelines and unequipped to guarantee results. Halse and Honey describe the 'disconnection' of the institution in which researchers are asked to put on a performance of a suite of textual competencies deemed necessary and desirable within the discourse's governing conditions: the ability to fill out the forms in the approved way, to deploy 'ethics-speak' as required, and to couch a research project in the language of scientific objectivity that resonates with many ethics review committees. (Halse & Honey, 2007, pp. 344–345)

They propose instead a model, informed by feminist practices, based on relational ethics of research in the 'real world'. This model, they claim,

> privileges relationality, care, vulnerability, and responsibility and asserts the importance of active concern for others and for community .... This does not mean uncritically discarding the moral principles of national ethics codes or creating a cavern where anarchy and ethical paralysis reign supreme. Rather, it means an ethical (re)orientation that arises from active engagement in the sort of critique that we have presented in our article: an unravelling of the taken-for-granted assumptions, ways of thinking, and technologies and tactics that shape us into particular sorts of researchers and configure research ethics in particular sorts of ways. (ibid., p. 349)

In their criticism of the dominant institutional frameworks, all of these image-based researchers reveal a frustration with a 'one size fits all' attitude towards ethics. They all emphasise the need to interrogate the assumptions and origins of the guidelines and policies under which researchers operate. The alternatives they advocate, moreover, all centre the researcher's own role as an autonomous, empathetic and integrity-seeking individual.

VISUAL DATA

In answer to the sort of criticisms outlined above, guidelines have been developed to inform image-based research more specifically. The chief example is the International Visual Sociology Association (IVSA) Code of Research Ethics and Guidelines (2009). This document defines the ethical and professional responsibilities of researchers, from multiple disciplines, working with image-based material. The IVSA Code was created in response to demand from university researchers and professional organisations (Papademas, 2009, p. 1). The Code clearly places the final responsibility and determination with the individual researcher, and tacitly supports a situated and flexible approach. It states that the guidelines in the Code

> should be used as guidelines for research activities in the conduct of professional work. As a guideline this document is intended to aid IVSA members to be aware of ethical issues in the research process, to encourage individual responsibility for ethical practice, and to provide a supportive document for visual researchers pursuing formal approvals from ethics review boards, academic institutions, and prospective sponsors. (Papademas, 2009)

The Code goes on to define a set of five general principles to guide the researcher – professional competence; integrity; professional and scientific responsibility; respect for people's rights, dignity, and diversity; and social responsibility – and a further set of ethical standards. These cover such areas as public communication and media presentation of research; and education and training support for a 'collegial network', alongside more widely documented ideals around subject consent and confidentiality (ibid.). The Code links institutional practice and individual integrity – resonant with the values-driven approach to ethics I advocate here.

Wiles et al.'s (2008) review 'Visual Ethics: Ethical Issues in Visual Research' outlines the key theories and frameworks guiding image-based researchers, and describes a discrete set of ethical concerns which researchers might encounter. The article goes on to suggest how researchers should deal with these concerns. Taking an anthropological and sociological perspective on visual research,

Wiles et al. make only glancing reference to image consumption and construction. Their focus is on gathering, producing, and exhibiting images rather than studying them, so their work considers principles of consent, confidentiality, and anonymity. The authors note other areas the researcher •

> needs to attend to: the internal meanings of an image; how it was produced; and how it is made meaningful by its viewers ... key issues to be considered by researchers are: i) the context in which the image is produced ii) the content of the image iii) the contexts and subjectivities through which the images are viewed. (Wiles et al., 2008, p. 33)

The authors regard these concerns about consumption and construction through the eyes of practitioners, however, concluding 'Visual researchers need to carefully consider these issues in order to take seriously the promises they have made to their study participants' (ibid., p. 34).

There are, of course, many more questions facing the image-based researcher, questions which seem to fall outside the scope of the existing guidelines and recommendations. When thinking about how to be an ethical scholar of images, for instance, we might ask:

- Who feels ownership over the images we use? Who actually owns them and how do we interact with them accordingly? How do we ethically use images when teaching or presenting our research?
- Are the images we use ethically produced or collected?
- What about artists who have themselves behaved unethically? How should we engage with the art they have produced?
- Why are some images seen as more valuable or useful than others?
- How do we select and use images in our research?
- How do we deal with the affective power of images? Do we have a responsibility, when we circulate images, to consider their affective power on others?
- What assumptions and biases do we bring to our research, and how do we acknowledge this in our work?

The dissemination of research outputs is another area of concern. In an institutional climate that demands researchers demonstrate 'Impact', what about the real impact on the communities which are subjects of, or subject to, our research? Open access requirements make published work more widely available, taking it from behind paywalls and library walls, but to date, there is uncertainty around some of the logistics and financing of open access, creating limitations and challenges for researchers. How should the findings of research about a subject or group of subjects be made available, and how should it be contextualised to make it truly accessible? The researcher has a duty of care to consider both the anticipated and the unanticipated effects of (mis)representation and exposure. Guidelines and regulations are not designed or best-equipped to define what sort of aftercare or amelioration can or should be provided.

When one begins to question the assumptions of previous generations of screen scholars, new approaches become not just possible, but necessary. We might think here of the challenges current scholars, such as Aaron Hunter (2016) in his work on New Hollywood Cinema, have mounted to the notion of single authorship. By looking anew at the contributions of cast and crew, seeing canonical works as products of collaborative authorship, these scholars do not just suggest new ways to analyse the meaning of the films. They also change the way we see the individual human subjects and workers whose labour has been undervalued, and interrogate a patriarchal view of the industry. Or, in another example, we might think of the work being done by scholars such as Daniela Treveri Gennari (2019), who looks at the social experiences of women attending Italian cinemas, considering how these film viewers experience the physical and cultural space of the cinema; placing value on emotional and interpersonal relationships and the voices of female spectators. The list goes on – but it is by broadening their views and questioning received wisdom that screen scholars are ethically approaching their research.

Elsewhere, and working in partnership with a historian, I have attempted to delineate some principles to guide the on-screen treatment and representation of historical events. This effort is underscored by the stance that filmed narratives do not just represent history; they actively participate in its construction (Carlsten & McGarry, 2015). Here, I would like to expand on these principles with reference to the ethics of film scholarship more broadly.

First, we should recognise that film-making and film-viewing are collaborative processes. Filmmakers, cast or subjects, crew, activists, and audience members are all stakeholders. These are all human beings with a role to play, and whose needs and responsibilities should be considered in an ethical evaluation of the work. How has the image before us, awaiting our analysis, been shaped by the input of these contributors? How does our intervention, as a scholar, impact on these individuals? When we consider that images are (at the point of production *and* reception) constructed from and interpreted through multiple, sometimes divergent perspectives, we open ourselves up to a more empathetic and complex understanding.

Second, screen scholars should work to broaden the notion of 'sources'. Contemporary image-based research is predicated on a shift away from written archives. Image-based researchers working with photographic and film archives give greater credence and exposure to oral testimony and the act of 'bearing witness' to history. Screen scholarship has seen a shift in the hierarchical ranking of archival material, with amateur film-making, home movies and 'found images' rising in status. As image-based researchers, we have an ethical responsibility to consider how the hierarchy of the archive informs or limits our work.

Likewise, when we reject the notion of a straightforward indexicality, and accept that the image bears a more complicated relationship to the real, we inevitably reject a valorisation of 'objectivity'. Image-based researchers must think in terms of multiple realities and multiple narratives. Instead of judging images based on their accuracy, we might think of their intent and their contract with the viewer, and of the polyvocality of images: whose voices are heard and whose visions are foregrounded?

This means of course that screen scholars have not just a pedagogical but an ethical obligation to consider the formal construction of the images we study.

We must not be fooled by claims to indexicality or by the false transparency of realism. Maintaining our commitment to formal analysis alongside textual or sociological analysis is an ethical act.

At the same time that we broaden our understanding of sources and stories, ethical scholarship requires that we consider the emotions that our research accesses, exploits, or engenders. Rather than striving to adopt an objective position, or demanding that of our subjects and texts, we might consider being more transparent and reflexive in our practices and our outputs. Rather than seeing emotion – our own or someone else's – as an obstacle to overcome, we might see it as an inherent and valuable resource for researchers.

As part of a more reflective practice, image-based scholars must consider the context in which images are produced. Increasingly, this means that researchers are attentive not just to national context, but regional, transnational, local and micro-contexts. Here, especially, learning from and listening to other stakeholders is essential to an ethical approach.

Screen scholarship has in recent years seen a greater emphasis on reception and audience. Along with this, an ethical approach might call for more attention to the discourse around films and to the paratexts and prosumers which extend, challenge, reframe, or rework images. How do we, as researchers, respond to or participate in these appropriations? As academics, most of us hope to be a part of that discourse: does this give us an additional burden of responsibility for how our contribution is used?

CONCLUSION

Clark's call for a situated ethics framework for visual research is a response to the inadequacies of existing institutional frameworks and ethics review bodies. Further criticising those frameworks and bodies, Halse and Honey point out that these often operate in an illusory way – by creating a set of boxes to be ticked off, they suggest a level of scrutiny and engagement which is not genuine. Moreover, these existing frameworks, they suggest, are doomed to failure, because they cannot account for the variety and change that is inherent in relationships between people.

> It is not possible to configure in advance and for ever more the sort of human beings that all research participants are or might be in the future or to assume that 'respect for persons' can be achieved through the independent actions of emotionally disengaged subjects. Nor can we assume that recourse to absolute moral rules will always be desired or relevant to all participants in all cases, contexts, or cultures… Our only resource is care and respect made possible in/through dialogical relations with others. When this basic moral principle is breached, it generates a moral void – the sort of spiritual emptiness that we experience when we renounce the call to a relationship with others. (Halse & Honey, 2007, pp. 345–346)

When we think about our relationships with others, we should not limit this either to the subjects of our research. Our relationships with colleagues, peers, and professional collaborators are important to our development as ethical scholars. Just as the 'tick boxes' of ethics committees are no substitute for a culture of ethical reflection, HR policies and workplace standards are no substitute for what Michael Kuhar and Dorthie Cross have called a 'collegial ethics'. Simply put, collegial ethics is the radical idea of being nice to one another in the workplace: being actively supportive in our language and behaviour, forgiving the mistakes of our coworkers, and showing courage in our work relations (Kuhar & Cross, 2013).

Our individual actions as researchers impact the integrity of our collaborations with others. Transparency around decision-making recognises people's autonomy and enables informed consent. Acting with empathy means not 'pulling up the drawbridge behind us', but actively working to improve conditions for those at earlier stages of their career, and supporting those in precarious 'casualised' labour. It requires generosity – giving time and lending voice to support one another. Listening to others, especially those whose lived experience does not resemble our own, creates space for reflection and self-inquiry. Regular discussion with other researchers, around the proverbial water cooler or around the world, enables debate and dialogue about how to address ethical concerns.

Emotional and moral engagement with our own research is not something to be avoided, but something to be examined. As Clark (2013) argues, it is in 'those moments of ethical anxiety when we are required to question the processes and assumptions underpinning our practices, that a visually-appropriate, situated ethical praxis can emerge' (p. 69). It is only through the on-going processes of personal reflection *and* dialogue with others that we can realise best practice in our own work. Rather than a 'Holy Grail' in the form of a set of guidelines and regulations governing our research practices, we would do better to pursue the ideal of a 'Round Table'. By developing non-hierarchical spaces (both physical and virtual) for the exchange of ideas and best practices, we can encourage greater rigour and ethical scrutiny. By bringing people together to share their experiences and perspectives, and to listen to the voices of other researchers, we can create opportunities for empathy. And by promoting an institutional culture that values the autonomous, ethical researcher, we can lay the groundwork for a holistic, values-driven approach to image-based research.

REFERENCES

Boddy, J., Neumann, T., Jennings, S., Morrow, V., Alderson, P., Rees, R., & Gibson, W. (2010). *The research ethics guidebook: A resource for social scientists*. Retrieved from http://www.ethicsguidebook.ac.uk

Carlsten, J. M., & McGarry, F. (Eds.). (2015). *Film, history and memory*. London: Palgrave Macmillan.

Clark, A. (2013). Haunted by images? Ethical moments and anxieties in visual research. *Methodological Innovations Online*, *8*(2), 68–81.

Dodd, S. (2019). What are photography ethics? Photography Ethics Centre. Retrieved from https://www.photoethics.org/photography-ethics

Halse, C., & Honey, A. (2007). Rethinking ethics review as institutional discourse. *Qualitative Inquiry*, *13*(3), 336–352.

Hunter, A. (2016). *Authoring Hal Ashby: The myth of the new hollywood auteur*. London: Bloomsbury.

Iphofen, R. (2013). *Research ethics in ethnography/anthropology*. Brussels: Ethics Unit B6, DG Research and Innovation of the European Commission.

Kuhar, M. J., & Cross, D. (2013). Collegial ethics: Supporting our colleagues, *Science and Engineering Ethics*, *19*, 677–684. doi:10.1007/s11948-012-9364-9

Neem, J. N. (2013). Making sense of the higher education debate. *Inside Higher Ed*. Retrieved from http://works.bepress.com/johann_neem/14/

Papademas, D. (2009). International visual sociology association code of research ethics and guidelines. *Visual Studies*, *24*(3), 250–257. doi:10.1080/14725860903309187

Pitt, P. (2014). "The project cannot be approved in its current form": Feminist visual research meets the human research ethics committee. *Australian Educational Researcher*, *41*(3), 311–325.

Treveri Gennari, D. (2019). Italian cinema audiences: A collaborative research project on cinema-going in 1950s Italy. Retrieved from http://italiancinemaaudiences.org/

Wiles, R., Prosser, J., Bagnoli, A., Clark, A., Davies, K., Holland, S., & Renold, E. (2008). *Visual ethics: Ethical issues in visual research* (p. 11). Southampton: ESRC National Centre for Research Methods. Retrieved from http://eprints.ncrm.ac.uk/421/1/MethodsReviewPaperNCRM-011.pdf

INDEX

Note: Page numbers followed by "*n*" indicate notes.

Printed and bound by CPI Group (UK) Ltd, Croydon, CR0 4YY

13/09/2026

14961059-0004